p94

DATE			

THE CALCIUM-REQUIREMENT COOKBOOK

THE CALCIUM-REQUIREMENT COOKBOOK

200 RECIPES THAT SUPPLY
NECESSARY CALCIUM-RICH FOODS TO
PREVENT THE BONE LOSS THAT OFTEN
BEGINS IN A WOMAN'S THIRTIES

JOANNE NESS and
GENELL SUBAK-SHARPE

M. Evans and Company, Inc.
New York

This book does not substitute for the medical advice and supervision of your personal physician. No change in diet should be undertaken except under the direction of a physician.

Library of Congress Cataloging in Publication Data

Ness, Joanne.
The calcium-requirement cookbook.

Includes index.
1. High-calcium diet—Recipes. 2. Low-calorie diet—Recipes. I. Subak-Sharpe, Genell J. II. Title.
RM237.56.N47 1985 641.5'632 85-1641

ISBN 0-87131-453-3

M. Evans and Company, Inc.
216 East 49 Street
New York, New York 10017

Design by Ronald F. Shey

Manufactured in the United States of America

9 8 7 6 5 4 3 2 1

To our parents and families

Contents

Acknowledgments

Few activities are performed singularly, and the writing of a cookbook is no exception. In this venture, Joanne Ness assumed the major task of developing and testing the recipes. From its very inception, friends and family of both the authors have generously offered their most treasured secrets (thanks, Mom and Aunt Evelyn).

A collective thanks to PJC friends. Special gratitude is extended to Sharon Altschuler, Sara Kaufman, Nancy Levin, Bina Rubanowitz, Ellene Newman, Ira and Dorothea Stokes. Thanks, too, to Barbara Turro, who painstakingly calculated the milligrams of calcium and calories for each recipe, and to Emily Paulsen and Gerald Subak-Sharpe, who attended to the many details necessary in producing a book. Our children, Jordan Ness and David, Sarah and Hope Subak-Sharpe, are commended for their taste-testing and sheer enthusiasm.

The Case for Calcium

Nutritionists often refer to calcium as the most important of the minerals necessary to sustain life. Yet even in this era of increasing nutrition awareness, most American adults consume diets that are woefully deficient in calcium. Many medical researchers are now convinced that calcium deficiency is a major cause of osteoporosis—brittle, porous bones—a disease that is epidemic among American women. A third of all women eventually develop this serious, disabling disease. Its complications contribute to up to 50,000 deaths a year, making it one of the most common killers of women over the age of 65. The direct medical costs are $3.8 billion a year and rising. But most experts are now convinced that these tragic statistics could be substantially improved *if women increased their calcium intake early in life.*

Although osteoporosis is the major consequence of long-term calcium deficiency, other common health problems are now being linked to it. Inadequate calcium may also contribute to periodontal disease—the breakdown in the bony structure supporting the teeth—which is the leading cause of tooth loss in this country. Recent research suggests that calcium deficiency also may contribute to high blood pressure, which is the leading cause of strokes and also a major factor in heart attacks and kidney disease. If so many serious, indeed life-threatening problems are linked to inadequate calcium, why has it taken so long to come to public attention? Unfortunately, the results of this chronic calcium deficiency are not immediately apparent, particularly in adults. Children who are deprived of calcium often develop rickets, a softening of the bones, leading to bowed legs and failure to grow normally. In this country, rickets is now a rare disorder because most children consume a diet high in calcium-rich foods, particularly milk—a food that many adults shun. It may take years for the chronic calcium imbalance to take its toll; typically, weak bones show up in middle age or later, often when it is too late to reverse the damage.

Contrary to popular belief, we never outgrow our need for calcium; in fact, a National Institutes of Health consensus statement on osteoporosis recommends that adult women should consume as much calcium as, or more than, the present Recommended Dietary Allowance for growing children.

To better appreciate the importance of calcium in the diet, let's briefly review its many functions. Calcium is the most abundant mineral in the body, accounting for about 2 percent of our total weight. The bones and teeth house 99 percent of this calcium; the remaining 1 percent circulates in the blood. Calcium, along with phosphorus, is essential in the building and maintenance of bones and teeth. But every cell in the body also requires minute amounts of calcium in order to function. Without calcium, the nerves cannot properly conduct impulses; muscles, including the heart, cannot contract; the brain will not function; blood will not clot properly.

To insure that each cell in the body has enough calcium, whenever the amount in the blood falls to a certain level, the mineral is released from the bones in a complex process that is still not fully understood. Calcium consumed in the diet is absorbed into the bloodstream from the small intestine. If there is enough calcium circulating in the blood to meet the needs of muscles, nerves, and other soft tissues, some of the excess is stored in the bones and some will be excreted, mostly in the urine, but also in the feces. As with so many other nutrients, the greater the need, the more efficient the absorption process. Thus, growing children and pregnant or nursing women, who have the greatest immediate need for large amounts of calcium, make more efficient use of dietary calcium than other people who still require calcium, but not so immediately.

Most of us think of the bones as static, solid structures. In reality, they are constantly undergoing change, with old bone cells being replaced by new. About every five years, the entire skeletal structure is renewed through this constant rebuilding. Bone renewal requires a steady supply of calcium to insure that the new bone is as strong and dense as that being replaced. But since the skeleton serves as a storehouse for the calcium needed by other body tissue, the mineral is constantly moving in and out of the bones. When calcium is released from the bones for use throughout the body, some may eventually be reabsorbed and some will be excreted. Thus, the body constantly requires new calcium from the diet simply to maintain its stores.

Unfortunately, not all of the calcium consumed in foods winds up working in the body because a number of factors affect the body's ability to absorb it. Vitamin D is essential to the complex process; a hormone derived from it via a mechanism involving the liver and kidneys stimulates intestinal absorption. Lactose, the sugar in milk, enhances calcium uptake. Some researchers also have found that absorption is enhanced by vitamin C, but hindered by a highly acidic diet. Absorption may be hindered by a high level of fat in the small intestine, although a small amount of fat seems to increase uptake.

Phosphorus, a companion mineral needed by the bones to make them hard, competes with calcium for uptake by the body. Only certain amounts of each mineral will be absorbed from the blood. A proper balance is maintained if the two minerals are consumed in about equal amounts. When there is an excessive amount of phosphorus, less calcium will be absorbed. Unfortunately, most of us eat much less calcium than phosphorus, which is found in such dietary favorites as meat, poultry, fish, and soft drinks. A high-protein diet also reduces calcium absorption.

Ironically, some foods that are high in calcium also contain oxalic acid, a substance that hinders calcium absorption. This acid is found in spinach, kale, mustard greens, rhubarb, chocolate, and certain other foods. Even though some of these foods contain large amounts of calcium, not all of it will be absorbed because of the oxalic acid. Similarly, a substance called phytic acid, which is found mostly in bran, hinders calcium absorption. Large amounts of tea also reduce calcium uptake.

Obviously, to get the maximum benefit from the calcium you eat, you must pay attention not only to how much calcium you consume, but also to combinations of foods that might keep your body from using it. The recipes in this cookbook are designed specifically to point you in the right direction. For example, the entrées and other dishes in this book do not contain red meat or poultry. We do not advocate that these foods be eliminated from the diet, but since these foods are high in phosphorus and low in calcium, we have not included them in the book. Plan your diet so that these high-phosphorus, low-calcium foods are not consumed in a meal in which you are trying to get a good amount of calcium.

Recent research indicates that our life-style also has an effect on calcium absorption. Cigarette smoking lowers the body's ability to absorb dietary calcium. The same is true of excessive alcohol consumption, defined as more than four or five drinks a day. Small amounts of calcium are lost through the skin during heavy sweating; marathon runners may need a bit more than more moderate exercisers. But people who don't exercise at all have a much greater calcium loss.

Numerous studies have found that inactivity, such as bed rest as a result of a fracture or illness, greatly increases the amount of calcium that is lost from the bones. A certain amount of mechanical stress is needed to promote remineralization of bone. This became apparent during the early days of subjecting astronauts to prolonged periods of weightlessness in preparing them for space flights. Without the normal stress of use against the force of gravity, the bones rapidly lost calcium and phosphorus. Thus, regular exercise is as important as adequate calcium in the diet in maintaining strong, healthy bones.

Finally, a number of hormones are important in maintaining calcium balance. Parathyroid hormone helps regulate the amount of calcium in the blood. Estrogen also appears to play a role in bone metabolism. Bone loss in women accelerates after menopause or removal of the ovaries, and studies have found that it may be slowed or perhaps even prevented by the prudent use of estrogen therapy.

SIGNS OF CALCIUM DEFICIENCY

As noted earlier, it takes years for the adverse effects of chronic calcium deficiency to become apparent. X-rays often will show a thinning of bones before symptoms such as pain, shortened stature, or fractures occur. During the first few decades of life, we constantly build bone mass, a process that peaks at about age 35. After that, bone mass declines. The more bone mass, the slower the decline. Men have about 30 percent more bone mass than women, and blacks have about 10 percent more than whites. This is one reason why women, particularly fine-boned whites or Orientals, are more vulnerable to osteoporosis than other population groups.

The jaws are among the first bones to become demineralized, a factor that may contribute to periodontal disease, even though bacteria are the major culprits in the breakdown of the gums and tissue supporting the teeth. Thus, loose teeth may well signal serious bone loss as well as periodontal disease.

Osteoporosis is one of the most common and serious consequences of long-term calcium deficiency. This disease afflicts an estimated 20 million Americans; one out of every three women will eventually develop it to varying degrees. Complications of osteoporosis account for 40,000 to 50,000 deaths a year, according to reports presented at a recent National Institutes of Health (NIH) conference on osteoporosis. Most of these deaths are precipitated by broken bones; osteoporosis causes 1.3 million fractures a year. For an older person, this often means hospitalization and surgery. Possible complications include pulmonary embolism, pneumonia, and other life-threatening consequences. And the forced inactivity leads to even greater bone loss and increasing disability. The total financial cost of osteoporosis is put at a staggering $3.8 billion a year. Of course, no dollar amount can be placed on the human suffering in terms of pain, disability, and loss of independence.

The vertebrae, which make up the spinal column, are particularly vulnerable to osteoporosis. As the vertebrae become demineralized, compression fractures occur. Sometimes the person is not aware that a vertebra has been fractured; more often, however, the fractures cause pain, distortion of the spine, and increasing disability. For example, as the spine becomes more compressed, height is lost and a characteristic curve or "widow's hump" develops in the upper back. The person can no longer stand fully erect and walking becomes more difficult.

The hips are still another common target of excessive bone loss. In fact, hip fractures are a major precipitating cause of death in older women. Very often, the hip fracture seems to be almost spontaneous; the victim's hip breaks, causing a fall, not vice versa. Or a minor bump, fall, or routine movement may result in a serious fracture. In any event, the fracture means hospitalization and, frequently, surgery. Among women sixty-five or older who suffer hip fractures, 20 percent eventually die of complications directly related to the injury.

The cause of osteoporosis is unknown, but increasing evidence indicates that inadequate calcium intake is at least partly to blame. Other factors include inactivity and the diminished estrogen that occurs with menopause. Women who undergo an early menopause, either because of surgical removal of the ovaries or an unusual natural menopause, develop osteoporosis at a much earlier age than women who undergo menopause between the ages of forty-five and fifty-two. Studies have found that the bone loss can be slowed or even prevented by increasing calcium intake, by maintaining a healthful level of physical activity, and, if appropriate, by taking low doses of estrogen supplements. Each case must be evaluated on an individual basis; increasingly, however, physicians and nutritionists agree that calcium is an important factor, both in preventing osteoporosis and in slowing the progression of the disease when it does occur.

HOW MUCH CALCIUM IS NEEDED?

The Recommended Dietary Allowance (RDA) of calcium for women is 800 milligrams, although the NIH consensus report on osteoporosis recommends that this be increased to at least 1000 milligrams a day. In keeping with this, the Food and Drug Administration has already increased its Recommended Daily Allowance (also known as RDA) of calcium to 1000 milligrams. More calcium—1200 milligrams a day—is recommended for growing children and for women during pregnancy and breast-feeding.

Some experts feel that the RDA for women is still too low. Many physicians are now urging women over the age of 35 or so to increase their calcium intake to 1000 to 1500 milligrams a day. But most adults do not consume even the 800 milligrams. According to the NIH report, "*the usual daily intake of elemental calcium in the United States, 450 to 550 mg., falls well below the National Research Council's RDA for calcium.*"

The report specifically urged that adult women consume 1000 milligrams of calcium a day, and that "postmenopausal women who are not treated with estrogen require about 1500 mg. daily for calcium balance." The report went on to note, "In some studies, high dietary calcium suppresses age-related bone loss and reduces the fracture rate in patients with osteoporosis. It seems likely that an increase in calcium intake to 1000 to 1500 mg. a day beginning well before the menopause will reduce the incidence of osteoporosis in postmenopausal women." But women were not the only subjects of the NIH report. "Increased calcium intake may prevent age-related bone loss in men as well."

Dr. Morris Notelovitz, Professor of Obstetrics and Gynecology at the University of Florida College of Medicine in Gainesville, has been one of the leaders in convincing his medical colleagues of the importance of helping women prevent osteoporosis. At a recent symposium in New York, Dr. Notelovitz stressed that "nutritional assessments of normal women reveal that *most women get about one-third to one-half of the daily required amount of calcium. . . .*"

Dr. Notelovitz, while stressing that we lack the tools to predict accurately who will develop osteoporosis, described the most likely targets: "White or Oriental women who are petite with fair complexions, have a family history of osteoporosis and have recently undergone a premature menopause will be at highest risk. . . . Added to this list, however, are a group of accelerating factors that frequently represent poor life-style habits. Lack of exercise, inadequate calcium intake and an excess of bone robbers, like caffeine, alcohol, smoking, acidic foodstuffs, animal protein (red meat), certain calcium binding foods (e.g., fiber and spinach) will all help increase calcium loss."

While many factors appear to be involved in maintaining healthy bones, there is little doubt that adequate calcium is the important key. Without it, the body simply cannot maintain strong, healthy bones. Since bone loss begins in the mid-thirties, it is best not to wait until the process begins to start preventive action. *Consuming adequate calcium should be a lifelong process.* You need it as a child to provide the calcium needed for growing bones and teeth. Pregnant and nursing women need extra calcium to provide for their babies, both before and after birth. And young women need to increase their calcium intake to help prevent the tooth and bone loss that is so common among middle-aged and older people.

High-Calcium Cooking Tips

Women trying to increase their dietary calcium often give up before even beginning: "I don't like drinking milk," "I'm watching my calories, and things like milk or cheese kill my diet," or "Milk and cheese are loaded with cholesterol, which isn't good for you either." It's true that you can fulfill your daily calcium requirement by drinking a quart of milk, which contains about 1165 milligrams of calcium. But this is by no means the only source of calcium, as you can see by referring to the chart of high-calcium foods in the appendix. This cookbook is designed to help you meet your daily calcium requirement by letting you select from a variety of foods. Variety not only makes your diet more interesting, it also gives you an opportunity to get other vital nutrients in addition to calcium. The following tips are a few of the ways you can increase dietary calcium and, at the same time, watch total calories, lower your cholesterol and fat intake, and improve your total nutrition.

MILK AND MILK PRODUCTS

- Switch to milk that has 1 percent fat, or better yet, to skim milk. The recipes in this book use skim milk, which has 90 calories per cup compared to 150 calories in a cup of whole milk.
- Be creative with nonfat dry milk. This is an inexpensive way to increase calcium content of foods with a rather modest rise in calories. A tablespoon of nonfat dry milk contains about 57 milligrams of calcium and 15 calories. Stirring a tablespoon of dry milk into a glass of skim milk will make it richer, and will give you 360 milligrams of calcium and 105 calories. Similarly, dry milk can be used in place of nondairy creamers (which are high in saturated fats) in coffee and can be added to many recipes without changing texture or taste to increase calcium content.
- Read labels. Key words such as "nonfat milk solids" often mean that the calcium content has been increased substantially. For example, 1 cup of low-fat plain yogurt with nonfat milk solids contains 452 milligrams of calcium and 127 calories, compared with 141 calories and 275 milligrams of calcium in a cup of whole-milk plain yogurt.
- Not all cheese is high in calcium. Most hard cheeses are high in calcium (and also high in fat, calories, and cholesterol). Part-skim or low-fat cheeses have fewer calories and less fat than regular cheese; in most, the calcium content is about the same or higher. Cream cheese has very little calcium and is not a good choice for people concerned with calories and cholesterol. Part-skim ricotta cheese, in contrast, is a good source of calcium—84 milligrams per ounce—and has less fat and cholesterol than cream cheese.

SARDINES, CANNED SALMON AND OTHER CANNED FISH

- Look for brands that have bones and do not remove them. Sardines and other canned fish are among the best sources of dietary calcium, but only if they contain bones. Many people buy the more expensive boneless varieties, and pass up a good opportunity to increase their calcium intake. Calories can be reduced by buying water-packed varieties. Or you can drain oil-packed fish to reduce the fat and calorie content. There also are a few low-sodium brands for people who are watching their salt intake. But read the labels carefully; many of the low-salt varieties are also boneless and oil-packed.

COMBINATIONS OF FOODS

- Avoid combining high-phosphorus and high-calcium foods (the reasons for this are explained earlier).
- Excessive use of bran may hinder calcium absorption. Fiber is an important element in a healthful diet, but too much bran, which contains phytic acid, interferes with absorption of calcium and certain other minerals. This will not be a problem if your diet provides a variety of foods, including vegetables and fruits, which are also good sources of fiber.
- Don't rely heavily on vegetables as your only sources of calcium. Some vegetables, particularly spinach, kale, mustard greens or chard, contain large amounts of calcium. But the oxalic acid in these foods binds to the calcium to form compounds that are not easily absorbed from the small intestine. Some of the calcium is absorbed, but not as much as from other foods that do not contain oxalic acid.

WHAT ABOUT SUPPLEMENTS?

Is it possible to meet your calcium requirements by taking supplements? Nutritionists agree that if you are unable to consume the recommended 1000 to 1500 milligrams of calcium a day from your diet, then you probably should consider taking supplements. However, as with other nutrients, dietary sources are preferable for several important reasons. Foods contain other necessary nutrients that you will not get from a calcium pill, so it is well worth making the effort to add calcium to your diet. Also, not all calcium supplements are readily absorbed. Calcium malate and calcium carbonate are considered the most absorbable; calcium gluconate, although not as readily absorbed as the other two, is an acceptable alternative. Since vitamin D is needed to absorb calcium, many supplements contain a combination of the two. Care must be taken, however, not to take too much vitamin D—in large amounts, it is highly toxic. It is also possible to take too much calcium; large amounts also can produce toxicity. In fact, an excess of calcium circulating in the blood (hypercalcemia) is a serious medical problem. This is usually a result of too much vitamin D, bone disorders that release too much calcium into the circulation, or hormonal disturbances. But excessive intake of calcium, either from supplements or calcium-containing antacids, also can produce hypercalcemia. People with a tendency to form kidney stones also may not be able to tolerate large amounts of calcium. As in virtually every aspect of good nutrition, too much is as bad as too little.

One further caution on calcium supplements: Lead contamination has been found in dolomite and bone meal, both types of calcium supplements that are widely available in health food or vitamin stores. Many nutritionists advise avoiding these supplements, especially if you are pregnant or breast-feeding, since both the fetus and young children are especially vulnerable to lead poisoning.

The wisest course in taking any nutritional supplement or in changing your diet is to check with your doctor first. Many physicians now urge women to increase dietary calcium and also to take a moderate calcium supplement to make up for any shortcomings in the diet. To find out if your diet is providing adequate calcium, you should keep a food diary for a few weeks. Write down everything you consume during the course of a day; make sure you include the amount. This will enable you or your doctor, dietitian or nutrition counselor to see if you are getting enough calcium (and other nutrients) from your diet. It will also reveal other eating patterns or habits that may need changing. You can then use the diary to more effectively increase your calcium intake and improve your overall eating habits.

ABOUT THE RECIPES

The recipes on the following pages have been developed to increase calcium intake, to taste good, and to add variety to your diet. Obviously, calcium is only one of many essential nutrients; the variety of foods used in this book will help you achieve a balanced diet. Throughout, we have attempted to follow sound nutritional and health principles without going overboard. We have elected to use margarine instead of butter; some cooks may prefer the latter, but in the interest of keeping cholesterol intake down, we are using margarine.

Some recipes are frankly for special occasions. Others can become everyday staples. The intent is to make you more conscious of calcium sources, and if given a choice, to pick the foods that contain this important mineral. For example, if you have a choice between an ordinary pickle or a pickled cauliflower, why not pick the latter? It tastes just as good and contains some calcium.

We also hope you will begin to look at your old favorite recipes in a new light. No cook lives on just one cookbook—we hope you make this one of your most frequent choices, but there's no reason to throw out all your others. But you can learn to add sources of calcium to your old recipes. Yogurt is a good, high-calcium substitute for sour cream; nonfat dry milk can be added to many recipes without altering the texture or taste; a high-calcium dressing can be used instead of the more ordinary and less interesting French or Italian. These are but a few examples of the kind of high-calcium cookery that can soon be almost automatic.

Appetizers and Snacks

The appetizers in this section are, for the most part, intended for the special occasions when friends drop in. But they also are loaded with good nutrition, including calcium. Some, such as Spanakopita, can double as entrées. Those that are high in calories are, of course, best eaten in moderation. One or two of the high-calorie appetizers can be served along with plates of fresh vegetables, such as broccoli or cauliflower, that are good sources of calcium, and yogurt or tofu dips.

STUFFED MUSHROOMS

- 18 large mushrooms
- 10 ounces frozen chopped spinach, thawed and drained
- 10 ounces frozen artichoke hearts, thawed
- 1 medium onion, chopped
- 2 garlic cloves, minced
- 1 tablespoon oil
- 1 tablespoon lemon juice
- ½ teaspoon salt
- ¼ teaspoon ground pepper
- 2 tablespoons dry bread crumbs
- 2 tablespoons grated Parmesan cheese

1. Wash mushrooms thoroughly. Remove and chop stems; set aside. Scoop out insides of mushrooms and place caps on a greased cookie sheet.

2. In a processor or blender, whirl spinach, artichokes, mushroom stems, onion and garlic until finely minced (not puréed).

3. Heat oil in a skillet and sauté minced vegetables for 3–5 minutes. Add lemon juice, salt, and pepper. Stir in bread crumbs.

4. Spoon mixture into mushrom caps and top with cheese. Bake 10–12 minutes in a 350° oven.

Yield: 6 servings.

Calcium: 98 mg. per serving.

Calories: 92 per serving.

MUSHROOM AND NUT PÂTÉ

- 2 tablespoons margarine
- 2 medium onions, chopped
- 3½ cups finely chopped mushrooms
- 1 cup ground hazelnuts
- 1½ cups cooked garbanzo beans, puréed
- 2 hard-boiled eggs, chopped
- Salt and pepper to taste
- 1 teaspoon dried thyme

1. Heat margarine in a large skillet. Sauté onions till limp, add mushrooms, and cook till softened. Add nuts and cook 2 minutes; cool.

2. Pour into bowl; mix in puréed beans and chopped eggs. Season with salt, pepper and thyme.

Yield: 3½ cups.

Calcium: 10 mg. per tablespoon.

Calories: 33 per tablespoon.

GOLDEN CURRIED NUTS

- 1 tablespoon margarine
- 1–2 teaspoons curry powder
- 1 tablespoon Worcestershire sauce
- 2 cups whole shelled almonds
- ½ cup golden raisins
- ½ cup sliced Calimyrna figs
- ½ cup shredded coconut

1. Preheat oven to 325°. Melt margarine in a skillet; stir in curry powder and Worcestershire sauce.

2. Stir in nuts and mix until well coated. Pour into a shallow baking dish and bake for 20 minutes until lightly browned; cool.

3. Mix nuts with raisins, figs, and coconut.

Yield: 3½ cups.

Calcium: 12 mg. per tablespoon.

Calories: 40 per tablespoon.

FIG AND HAZELNUT SPREAD

½ cup chopped figs
¼ cup water
⅓ cup ground hazelnuts
1 cup margarine, softened

1. Cook figs in the water. Purée in a blender, adding just enough water to keep machine running.

2. Brown hazelnuts in 2 tablespoons of the margarine.

3. Vigorously stir fig purée into the remaining margarine. When smooth, stir in hazelnuts. Serve at room temperature.

Yield: 2 cups.

Calcium: 7 mg. per tablespoon.

Calories: 65 per tablespoon.

Note: This is marvelous over pancakes or waffles, or as a spread for bread or toast. Store up to 2 weeks in the refrigerator.

EGG AND TOFU SPREAD

1 tablespoon oil
1 medium onion, chopped
½ cup sliced fresh mushrooms
1 cup bean sprouts
8-ounce cake firm tofu, drained and finely chopped
Salt, pepper, and paprika to taste
2 hard-boiled eggs, chopped
2 tablespoons mayonnaise

1. Heat oil in a skillet, add onion and sauté until lightly browned.

2. Add mushrooms and cook until tender.

3. Stir in sprouts and cook 1 minute; add tofu and cook until water is absorbed. Season with salt, pepper, and paprika.

4. Combine eggs with mayonnaise; add to tofu mixture.

Yield: 1¾ cups.

Calcium: 15 mg. per tablespoon.

Calories: 27 per tablespoon.

Note: This is a marvelous spread on crackers or bread. Best served at room temperature.

FRUITED CHEESE SPREAD

16 ounces low-fat plain yogurt
½ cup sliced almonds
¼ cup chopped figs
¼ cup shredded coconut (optional)
2 tablespoons maple syrup or honey

1. Empty yogurt into a double layer of cheesecloth. Hold ends together and close tightly with string or a rubber band. Attach to a wooden spoon or other implement and suspend over a bowl or jar to collect the liquid from the yogurt. Leave in refrigerator for 24 hours.

2. Remove yogurt from cheesecloth. It should be the consistency of soft cream cheese.

3. Add remaining ingredients. Use as a spread on bread or crackers.

Yield: 1¼ cups.

Calcium: 55 mg. per tablespoon.

Calories: 45 per tablespoon (includes coconut).

POTTED HERB CHEESE

3 cups grated Cheddar or other hard cheese
2 tablespoons margarine, melted
2 tablespoons dry sherry
2 tablespoons skim milk
1 tablespoon chopped parsley
1 teaspoon dried tarragon
1 teaspoon dry mustard
1 teaspoon dried marjoram
1 teaspoon dried basil

1. Combine all the ingredients in the top of a double-boiler. Heat and stir constantly until cheese melts.

2. Pour into a jar, cover, and chill. Soften slightly before serving.

3. Serve with crackers or bread.

Yield: 2 cups.

Calcium: 48 mg. per tablespoon.

Calories: 30 per tablespoon.

KIPPER CANAPÉS

3½-ounce can kippers
½ lemon
2 tablespoons mayonnaise
1 teaspoon Dijon mustard
Hot pepper sauce (e.g., Tabasco) to taste
Crackers or toast rounds*
Garnishes: pimiento, olives, or paprika

1. Mash kippers. Do not remove bones.

2. Sprinkle with juice from ½ lemon.

3. Stir in mayonnaise, mustard and hot pepper sauce.

4. Spread on crackers or toast rounds. Garnish.

Yield: 12 canapés.

Calcium: 9 mg. per serving of filling.

Calories: 36 per serving of filling.

* Not included in calorie or calcium calculations.

BROILED CLAM CANAPÉS

- 1 cup minced clams
- 1 cup shredded Monterey Jack cheese
- ½ cup chopped celery
- ¼ cup low-fat plain yogurt
- 2 tablespoons pickle relish
- 1 tablespoon grated onion
- Sprinkling of paprika
- 25 crackers or pieces of melba toast

1. Combine clams, cheese, celery, yogurt, relish and onion.

2. Spread on crackers. Sprinkle with paprika and broil until cheese melts. Serve immediately.

Yield: 25 canapés.

Calcium: 34 mg. per canapé.

Calories: 36 per canapé.

SPANAKOPITA

- 20 ounces fresh spinach
- 3 tablespoons olive oil
- 1 cup chopped scallions
- 3/4 pound feta cheese, chopped
- 1/2 cup low-fat cottage cheese or ricotta
- 4 eggs, beaten
- 1/2 cup chopped fresh parsley
- 2 tablespoons fresh dill, or 2 teaspoons dried
- Sprinkling of black pepper
- 1/2 pound margarine, melted
- 1 pound filo dough, defrosted
- Sprinkling of sesame seeds (approximately 2 tablespoons)

1. Wash spinach, pat dry, and chop.

2. Sauté spinach in oil until tender. Add scallions and sauté 1 minute longer. Drain off any excess liquid.

3. In a bowl, mix cheeses, eggs, parsley, dill, and pepper. Stir in spinach-scallion mixture.

4. Spread some melted margarine in the bottom of a 9x13-inch baking dish. Carefully place a sheet of filo dough in the dish. Cover remaining sheets with a damp cloth. Brush dough in pan with margarine. Repeat this process with 9 sheets.

5. Spread spinach-cheese mixture on dough. Cover with 10 more sheets of filo dough, brushing each sheet with margarine as it is added.

6. Brush top with margarine. Sprinkle with sesame seeds. Bake 50–60 minutes at 350° until golden brown. Serve hot.

Yield: 8 entrée servings, 16 appetizer servings.

Calcium: 346 mg. per entrée serving; 173 mg. per appetizer.

Calories: 590 per entrée serving; 295 per appetizer.

SPINACH PUFFS

- 1 small onion, chopped
- 2 garlic cloves, minced
- 1 tablespoon margarine
- 10 ounces frozen chopped spinach, thawed and drained
- 1 tablespoon chopped fresh parsley
- ½ teaspoon salt
- ¼ teaspoon ground pepper
- 8 ounces farmer or hoop cheese
- 3 ounces Cheddar cheese, grated
- 8 frozen puff pastry squares, thawed

1. Sauté onion and garlic in margarine until limp. Add spinach and sauté 5 minutes longer. Stir in parsley, salt, and pepper.

2. Remove from heat and mix in cheeses.

3. Place 1½ tablespoons filling in the center of each pastry square. Fold and seal with water.

4. Place in lightly greased baking dish. Pierce tops with a fork. Bake in a 400° oven for 20–25 minutes.

Yield: 8 puffs.

Calcium: 223 mg. per puff.

Calories: 158 per puff.

HUMMUS

- 16-ounce can garbanzo beans
- 4 tablespoons tahini (sesame seed paste)
- 4–5 tablespoons fresh lemon juice
- 3 garlic cloves, minced
- Dash salt
- 1 tablespoon sesame seeds
- 1 tablespoon olive oil
- Paprika

1. Drain garbanzo beans and reserve liquid.

2. Place beans, tahini, lemon juice, garlic, salt, and sesame seeds in a blender. Purée, adding enough reserved bean liquid to keep machine running.

3. Pour into a shallow serving dish. Cover and chill.

4. Just before serving, sprinkle with olive oil and paprika.

5. Serve with crackers or in pieces of pita bread.

Yield: 2 cups.

Calcium: 16 mg. per tablespoon.

Calories: 37 per tablespoon.

PARSLEY SPINACH HUMMUS

- 1 cup chopped fresh spinach leaves
- ⅔ cup chopped scallions
- ⅔ cup chopped parsley
- ⅓ cup oil
- 16-ounce can garbanzo beans
- ½ cup tahini (sesame paste)
- 2 garlic cloves, minced
- Juice of 1 lemon
- Salt to taste
- Salad greens (optional)*

1. Place spinach, scallions, parsley, and oil in a blender or food processor and whirl until very finely minced. Pour into a large bowl.

2. Drain garbanzo beans, but reserve liquid. Place beans, tahini, garlic, and lemon juice in blender or processor. Add enough reserved bean liquid to keep machine running; purée beans.
3. Spoon onto greens and season with salt.

Yield: 2½–3 cups.

Calcium: 88 mg. per ¼ cup serving.

Calories: 154 per ¼ cup serving.

* Not included in calorie and calcium calculations.

EGGPLANT DIP

1 medium eggplant
1 small onion, grated
1 cup low-fat plain yogurt
⅓ cup sunflower seeds
1 teaspoon Italian seasoning
¾ teaspoon salt
¼ teaspoon ground pepper
Hot pepper sauce (e.g., Tabasco)

1. Bake eggplant in a 450° oven for 30–40 minutes, or until skin splits and pulp is softened. Remove from oven and cool. Peel and mash eggplant.
2. Place pulp in a blender or food processor along with remaining ingredients; purée.
3. Place in a bowl and season with hot pepper sauce. Serve with crackers or sliced fresh vegetables.

Yield: 2 cups.

Calcium: 19 mg. per tablespoon.

Calories: 24 per tablespoon.

MEXICAN CHEESE DIP

- 16 ounces low-fat plain yogurt
- 2 tablespoons minced pimiento
- 2 tablespoons minced fresh mint or cilantro
- 1 teaspoon minced dried onion
- 1 teaspoon chili powder
- Salt and pepper to taste

1. Follow directions for making yogurt cheese in steps 1 and 2 of Fruited Cheese Spread (see Index).

2. Add remaining ingredients. Use as a dip for taco chips or crackers.

Yield: $1\frac{1}{4}$ cups.

Calcium: 42 mg. per tablespoon.

Calories: 15 per tablespoon.

Soups

Some of the soups in this section are almost a meal in themselves; others are a low-calorie, high-calcium way to begin a meal. Serve the Black Bean Soup with a salad and you have a hearty lunch or a light supper. The same is true for the Curried Lentil Soup and the New England–Style Chowder. The Tomato-Yogurt Soup can be prepared in minutes and is ideal for a hot summer day. The calcium content of the creamed soups can be increased by adding a tablespoon of nonfat dry milk.

BLACK BEAN SOUP

1 pound dried black beans
2 quarts water
1 tablespoon salt
2 onions, chopped
5 garlic cloves, minced
1 tablespoon oil
1 teaspoon cumin
2 teaspoons oregano
1 teaspoon black pepper
Juice of 1 lime
1½ cups cooked rice (optional)
3 tomatoes, diced
3 scallions, chopped

1. Rinse beans. Place in a large pot, pour in cold water to cover, and soak overnight.

2. Bring 2 quarts water to a boil with salt. Drain beans and add to pot; simmer till tender (1½–2 hours).

3. While beans are cooking, sauté onion and garlic in oil till lightly browned. Stir in cumin, oregano and pepper. Add to cooked beans and simmer 30 minutes. Stir in lime juice.

4. Mix the rice with the tomatoes and scallions. Divide among 6 bowls. Pour in soup.

Yield: 6 servings.

Calcium: 151 mg. per serving.

Calories: 366 per serving (includes rice).

CURRIED LENTIL SOUP

1 tablespoon oil
1 onion, chopped
2 garlic cloves, minced
1 teaspoon minced fresh ginger
1 tablespoon curry powder
1 teaspoon each: salt, cumin, turmeric
6 cups water
2 chicken, beef or vegetable bouillon cubes
1½ cups dried lentils
Marinated Rice (recipe follows, optional)

1. In a large pot, heat oil and sauté onion, garlic and ginger until softened. Stir in spices and salt.

2. Add water, bouillon cubes, and lentils. Bring to a boil, reduce heat, cover, and simmer 1½ hours. Stir occasionally, adding more water if necessary.

3. If desired, place ¼ cup rice in each bowl, then pour in soup.

Yield: 6 servings.

Calcium: 38 mg. per serving. (Add 8 mg. if served with rice.)

Calories: 211 per serving. (Add 101 calories if served with rice.)

MARINATED RICE FOR LENTIL SOUP

1½ cups cooked rice
2 tablespoons oil
1 tablespoon vinegar
1 scallion, chopped
½ teaspoon dried basil
Dash of salt

1. Combine all ingredients.

2. Let stand at room temperature for 2 hours.

Yield: 6 servings.

TOMATO-YOGURT SOUP

35-ounce can peeled plum tomatoes
¾ teaspoon curry powder
½ teaspoon garlic powder
Juice of 1 lime
2 cups low-fat plain yogurt
Scallions for garnish

1. Drain tomatoes, but reserve liquid.

(*recipe continues*)

2. Blend tomatoes in a processor or blender to a coarsely chopped stage.

3. Stir in spices and lime juice.

4. Add yogurt and enough reserved tomato juice until a desired consistency is achieved.

5. Chill; taste and add more seasonings if desired.

6. Pour into bowls. Garnish with sliced scallions.

Yield: 6 servings.

Calcium: 150 mg. per serving.

Calories: 86 per serving.

TWO-STEP MINESTRONE

2 medium onions, sliced
4 garlic cloves, minced
1 tablespoon oil
4 plum tomatoes, chopped
2 cups tomato sauce
2 carrots, sliced
2 medium zucchini, sliced
2 cups shredded cabbage
1 cup (3 ounces) shells or other small pasta
1 cup cooked navy beans
1 cup cooked kidney beans
2 teaspoons Italian seasoning

1. Sauté onion and garlic in oil until transparent.

2. Add remaining ingredients, cover with water, and simmer 1 hour.

Yield: 8 servings.

Calcium: 61 mg. per serving.

Calories: 161 per serving.

Note: Best if made earlier in day and refrigerated, then reheated.

CREAM OF KALE SOUP

1 pound fresh kale
2 tablespoons peanut or vegetable oil
1 medium onion, chopped
¾ teaspoon powdered ginger, or 2 teaspoons minced fresh ginger
3 cups vegetable, mushroom, or onion stock
1 tablespoon Worcestershire sauce
1 cup nonfat dry milk
1 cup low-fat plain yogurt
2 tablespoons finely chopped scallions

1. Wash kale very well. Discard tough outer leaves and remove inner vein. Cut into bite-size pieces and set aside.

2. In a large pot, heat oil. Sauté onion until softened. Add ginger and cook 1 minute, stirring constantly.

3. Add kale and sauté until soft (add a little stock, if necessary).

4. Add stock and Worcestershire sauce; cook 10 minutes. Remove from heat and cool.

5. Stir in milk and yogurt; heat but do not allow to boil.

6. Garnish with chopped scallion.

Yield: 6 servings.

Calcium: 321 mg. per serving.

Calories: 139 per serving.

PARSLEY SOUP

1 cup coarsely chopped parsley
1½ cups water or chicken or vegetable broth
1 tablespoon margarine
1 tablespoon flour
1 cup skim milk
1 teaspoon salt

(*recipe continues*)

1. Combine parsley and water and bring to a boil in a large pot. Lower heat, cover, and simmer 7 minutes.

2. Pour mixture into a blender or food processor; whirl for 1 minute.

3. Add remaining ingredients and blend again. Return to pot and simmer 5 minutes. Serve hot.

Yield: 2 servings.

Calcium: 312 mg. per serving.

Calories: 113 per serving.

CREAM OF VEGETABLE SOUP

2 tablespoons margarine
1 medium onion, chopped
2 garlic cloves, minced
3 medium carrots, sliced
2 celery ribs, chopped
1 medium potato, peeled and chopped
¼ cup chopped parsley
2 cups vegetable stock or water
½ teaspoon salt
¼ teaspoon pepper
2 cups skim milk

1. In a large pot, melt the margarine and sauté onions and garlic until limp.

2. Stir in carrots, celery, potato, and parsley; sauté 10 minutes, adding a little water if necessary to prevent sticking.

3. Add stock, salt, and pepper; simmer over low heat for 45 minutes.

4. Purée half the mixture in a blender.

5. Return purée to pot; add skim milk and simmer until heated.

Yield: 6 servings.

Calcium: 160 mg. per serving.

Calories: 119 per serving.

CHILLED STRING BEAN SOUP

1 medium onion, chopped
1 tablespoon margarine
4 cups trimmed and sliced string beans
4 cups water
2 medium potatoes, cubed
1/4 cup chopped parsley
3/4 teaspoon salt
1/8 teaspoon pepper
1 1/2 cups skim milk
1 cup low-fat plain yogurt

1. In a large saucepan, sauté onion in margarine until limp. Add string beans, water, potatoes, parsley, salt, and pepper. Simmer 20–30 minutes. Cool to room temperature.

2. Add milk and yogurt. Chill and serve.

Yield: 6 generous servings.

Calcium: 222 mg. per serving.

Calories: 129 per serving.

BORSCHT WITH SPINACH

1 bunch beets (approximately 1 pound), peeled and grated
10 ounces spinach, washed and stems removed
1 1/2 quarts water
1/3 cup lemon juice
1 1/2 tablespoons sugar
1 teaspoon salt

1. In a large saucepan, combine all ingredients and simmer for 1 hour; chill.

2. Pour into bowls and top with dollops of low-fat plain yogurt, if desired.

Yield: 4 servings.

Calcium: 87 mg. per serving.

Calories: 90 per serving.

POTATO SPINACH SOUP

1 tablespoon margarine
1 cup chopped onion
10 ounces frozen chopped spinach
4½ cups water
1 pound new potatoes, peeled and cubed
½ pound sweet potatoes, peeled and cubed
½ teaspoon nutmeg
½ teaspoon salt
Pepper to taste
2½ cups buttermilk (from skim milk) or skim milk

1. Melt magarine in soup pot and sauté onion till lightly browned. Add spinach and cook until coated with margarine.

2. Add water, potatoes, nutmeg, salt and pepper. Cook until potatoes are tender.

3. Add milk and simmer 10 minutes longer. Refrigerate.

Yield: 8 servings.

Calcium: 146 mg. per serving.

Calories: 134 per serving.

Note: For a thick, smooth soup, purée the soup, one half at a time, in a blender or processor.

BROCCOLI BUTTERMILK SOUP

1 small bunch broccoli
½ cup diced green pepper
¼ cup chopped onion
1¼ cups water
½ cup buttermilk (from skim milk)
1 cup evaporated skim milk
⅛ teaspoon curry powder
1 teaspoon salt

1. Remove and discard tough parts of broccoli stems, then slice broccoli vertically into thin strips. You should have 2 cups.

2. Place broccoli in a large pot with green pepper, onion, and water. Over low heat, bring to a simmer, cover, and cook 10 minutes.

3. Pour into a blender or food processor and whirl until smooth. Pour mixture back into pot.

4. Add buttermilk, evaporated skim milk, curry powder, and salt. Heat thoroughly without boiling. Serve hot or chilled.

Yield: 3 servings.

Calcium: 409 mg. per serving.

Calories: 112 per serving.

BUTTERMILK FRUIT SOUP

- 1¼ cups cleaned berries (blueberries, strawberries, or blackberries)
- 1 large banana, peeled and sliced
- 2 cups buttermilk (from skim milk)
- ⅓ cup sugar
- Juice and grated rind of ½ lemon
- 1 teaspoon almond or vanilla extract
- Garnish: whole berries or fresh mint leaves*

1. Place berries, banana, buttermilk and sugar in container of blender or food processor and purée fruit.

2. Stir in lemon rind, juice, and extract.

3. Chill, garnish, and serve.

Yield: 4 servings.

Calcium: 179 mg. per serving.

Calories: 264 per serving.

* Not included in calorie or calcium calculations.

SEAFOOD SOUP WITH COUSCOUS

3 garlic cloves, minced
1 medium onion, sliced
1 tablespoon oil
1 celery rib, sliced
1 carrot, thinly sliced
6 cups water
1 pound fresh tomatoes, chopped
10 ounces frozen chopped spinach, thawed and drained
1 fish head, wrapped in cheesecloth
¾ pound mussels, scrubbed and debearded
¾ pound shrimp, peeled
Couscous (recipe follows)
Grated Parmesan cheese (optional)

1. Sauté garlic and onion in oil until lightly browned. Stir in celery and carrot; cook 2 minutes.

2. Add water, tomatoes, spinach, and fish head; simmer 30 minutes. Remove fish head.

3. Stir in mussels and shrimp; simmer gently 10 minutes.

4. Place one spoonful of couscous in each bowl. Ladle soup over and serve with cheese, if desired.

Yield: 8 servings

Calcium: 144 mg. per serving. (Add 69 if 1 tablespoon Parmesan is added.)

Calories: 197 per serving. (Add 23 if 1 tablespoon Parmesan is added.)

COUSCOUS FOR SOUP

1 cup couscous
2 tablespoons oil
½ teaspoon oregano
¼ teaspoon cinnamon

1. Brown couscous in oil, stirring constantly. Stir in spices.

2. Add amount of water suggested on package (this will vary with the type of couscous used) and cook according to package directions.

QUICK AND LIGHT POTAGE DE CRESSON

1 pound boiling potatoes
1 bunch watercress
2 quarts salted water
1 small onion, grated
3½ cups skim milk
Dash of sugar (optional)*
Pepper to taste
2 tablespoons margarine (optional)*

1. Peel, cut in half, and thinly slice potatoes. Wash the watercress and separate leaves from stems. Reserve some leaves for garnish. Coarsely chop remaining leaves; finely chop stems.

2. Bring water to a boil; add potatoes and watercress, boiling until potatoes are tender. Drain.

3. In a blender, purée the cooked vegetables and onion with 1½ cups of the milk.

4. Return purée to saucepan; add remaining milk, sugar, pepper, and margarine if a richer soup is desired. Reheat.

5. Pour into bowls. Garnish with watercress leaves.

Yield: 6 servings.

Calcium: 192 mg. per serving.

Calories: 126 per serving.

* Not included in calorie and calcium calculations.

NEW ENGLAND-STYLE CHOWDER

1 tablespoon margarine
1 cup minced onion
½ cup chopped celery
½ cup sliced mushrooms
½ cup chopped carrots
4 cups skim milk
2 large potatoes, diced
Salt and white pepper to taste
¼ cup minced parsley
2 bay leaves
½ teaspoon dried thyme
1 pound firm-fleshed fish fillets, cut into bite-size pieces, or 1 pint shucked clams

1. Melt margarine in a 5-quart pot; add vegetables and sauté till tender, adding a bit of water to prevent sticking.

2. Add milk, potatoes, salt, pepper, and herbs; simmer until potatoes are tender.

3. Add fish and simmer 10 minutes longer.

Yield: 6 servings.

Calcium: 247 mg. per serving.

Calories: 226 per serving.

ENTRÉES

These entrées are intended specifically to increase calcium intake. You will note that recipes containing meat and poultry have not been included. This does not mean that these foods should not be part of your diet; they are important sources of protein and other essential nutrients. But most of us already consume more than enough of these foods. Since meat and poultry are high in phosphorus and low in calcium, consuming an excess of these foods adds to a calcium imbalance because the phosphorus hinders calcium absorption, unless the two are consumed in about equal amounts. Therefore, we have concentrated on presenting recipes that provide a balance of the two minerals, or that offer large amounts of calcium to compensate for the meals in which you will be consuming red meat or other high-phosphorus foods.

Many of the entrée recipes contain spinach and other foods that contain good amounts of calcium. But some of these foods also contain oxalic acid, which hinders calcium absorption. But because some calcium will be absorbed, and because the foods contain other important nutrients, we think they are worthy of inclusion. Just remember they are not as calcium-rich as dairy products.

Remember, too, that calories can be reduced by cutting portion size or further reducing the amount of margarine or oil used in preparation. If you use a nonstick pan for sautéing, for example, you might use a pan spray instead of oil. Nuts and seeds have been added to many of the recipes to give extra calcium and flavor; in many recipes, however, these are optional ingredients. People who are counting calories can skip the nuts or seeds and still get a good amount of calcium.

KIPPERS WITH SAUTÉED VEGETABLES

1 tablespoon oil
1 medium green pepper, cut into thin strips
2 garlic cloves, minced
1 cup shredded cabbage
3½-ounce can kippers (kipper snacks)
1 scallion, chopped
Juice of ½ lemon
Generous sprinkling of pepper

1. Heat oil in a skillet. Sauté green pepper strips and garlic until slightly softened.

2. Add shredded cabbage and cook until softened.

3. Cut kippers into bite-size pieces. Add kippers and scallion to skillet; heat.

4. Sprinkle with lemon juice and pepper.

Yield: 1 serving.

Calcium: 214 mg. per serving.

Calories: 399 per serving.

BAKED SALMON AND PASTA

1 medium onion, chopped
3 garlic cloves, minced
1 tablespoon margarine
1 cup sliced mushrooms
2 cups thinly sliced zucchini
1 cup tomato sauce
1 teaspoon oregano
Sprinkling of ground pepper
3 ounces American or mozzarella cheese, shredded
15½-ounce can salmon in chunks
3 cups cooked pasta (shells or spirals)
Parmesan cheese, grated (optional)*

1. Sauté onion and garlic in margarine until limp. Add mushrooms, then zucchini. Cook until softened, stirring frequently.

2. Add tomato sauce, oregano, and pepper and cook for 15 minutes.

3. Add cheese, salmon, and pasta; continue cooking until cheese is melted and salmon is heated. Sprinkle with Parmesan, if desired.

Yield: 6 servings.

Calcium: 221 mg. per serving.

Calories: 352 per serving.

* Not included in calorie and calcium calculations; 1 tablespoon contains 69 mg. calcium; 23 calories.

SALMON PATTIES WITH YOGURT-DILL SAUCE

15½-ounce can salmon, drained
½ cup soft bread crumbs
¼ cup minced onion
1 egg, beaten
1 tablespoon chopped fresh parsley
1 tablespoon lemon juice
1 teaspoon Worcestershire sauce
¼ teaspoon garlic powder
Yogurt-Dill Sauce (see Index)

1. Combine all ingredients and form into patties.
2. Brown on each side in a lightly greased skillet.
3. Serve with Yogurt-Dill sauce.

Yield: 4 servings.

Calcium: 203 mg. per serving.

Calories: 300 per serving.

SALMON QUICHE

1 tablespoon margarine
1 medium onion, chopped
10 ounces frozen chopped spinach, thawed and drained
½ teaspoon salt
½ teaspoon garlic powder
⅛ teaspoon ground pepper
15½-ounce can salmon, drained
3 eggs, beaten
1 cup buttermilk (from skim milk)
8-inch pie shell, unbaked
¼ pound Swiss or Cheddar cheese, grated

1. Sauté onion in margarine till limp. Add spinach, salt, garlic powder, and pepper; cook until spinach is well-coated with oil. Preheat oven to 375°.

2. In a mixing bowl, flake salmon into pieces; stir in eggs and buttermilk.

3. Place salmon mixture in pie shell. Top with spinach-onion mixture. Sprinkle cheese on top.

4. Bake for 45 minutes.

Yield: 6 servings.

Calcium: 317 mg. per serving.

Calories: 456 per serving.

SHRIMP WITH ORANGE SAUCE

1 tablespoon margarine
1/4 cup finely chopped onion
1 tablespoon plus 1 teaspoon flour
3/4 cup skim milk
1 teaspoon freshly grated orange rind
1/8 teaspoon salt
Dash of cayenne pepper
1 pound boiled shrimp

1. Melt margarine in a saucepan and sauté onion until limp. Stir in flour and cook 2 minutes, stirring constantly.

2. Stir in milk, orange rind, salt, and cayenne; cook until thickened.

3. Stir in shrimp. Serve over cooked rice.

Yield: 4 servings.

Calcium: 193 mg. per serving (rice not included).

Calories: 189 per serving (rice not included).

SHRIMP MEDITERRANEAN

1 cup tomato sauce
2 medium tomatoes, chopped
½ medium onion, chopped
1 teaspoon dried oregano
1¼ pounds shrimp, peeled and deveined
½ cup feta cheese, crumbled

1. Simmer sauce, tomatoes, onion, and oregano for 20 minutes.

2. Preheat oven to 350°.

3. Place shrimp in a single layer in a greased baking dish.

4. Pour sauce over shrimp. Sprinkle with cheese and bake 20–25 minutes.

Yield: 4 servings.

Calcium: 282 mg. per serving.

Calories: 256 per serving.

Note: Grated Parmesan cheese may be substituted for feta cheese. If so, one serving would contain 247 mg. calcium and 209 calories.

SHRIMP INDONESIAN STYLE

2 pounds shrimp, peeled and deveined
Oil
½ cup ground hazelnuts or almonds, toasted in oven
3 tablespoons soy sauce
2 tablespoons molasses
8-ounce can crushed pineapple with juice
2 garlic cloves, minced
½ teaspoon crushed red pepper

1. Thread shrimp on skewers or place directly in baking dish. Brush with oil.

2. Prepare sauce by combining remaining ingredients and simmering 10 minutes. Baste shrimp with sauce.

3. Broil on one side, turn and baste other side; broil second side.

Yield: 8 servings.

Calcium: 169 mg. per serving.

Calories: 217 per serving.

SHRIMP CURRY

- 1¼ pounds shrimp, peeled and deveined
- 1½ cups sliced cauliflower florets
- 1½ cups sliced carrots
- 2 tablespoons oil
- 1 medium onion, chopped
- 3 garlic cloves, minced
- 2 tablespoons flour
- 2 teaspoons curry powder
- ¾ teaspoon salt
- ¼ teaspoon crushed red pepper
- 1½ cups fish or chicken stock, simmering
- 1 cup frozen peas
- 1 cup low-fat plain yogurt

1. In a large pot, steam shrimp, cauliflower, and carrots until almost tender; set aside.

2. In a large skillet, heat oil and sauté onion and garlic until soft. Stir in flour, curry powder, salt, and pepper. Cook for 2 minutes, stirring constantly.

3. Pour in stock and peas. Simmer 10 minutes, stirring, until thickened. Add the shrimp, cauliflower, carrots and yogurt and heat thoroughly; do not boil.

Yield: 4 servings.

Calcium: 242 mg. per serving.

Calories: 324 per serving.

Note: This is best when prepared a day ahead, refrigerated, then reheated.

SWEET AND SOUR GLAZED SHRIMP

1 tablespoon sesame seeds
1 tablespoon margarine
1/2 cup thinly sliced celery
1/2 cup chopped onion
1/4 cup chopped dried apricots, plumped in water
2 tablespoons apricot preserves
2 tablespoons water
1 pound shrimp, shelled and deveined

1. Sauté seeds in margarine, stirring until well coated. Stir in celery and onion; sauté until lightly browned.

2. Drain apricots and add to pan along with preserves and water; cook 5 minutes.

3. Stir in shrimp and sauté 10 minutes until tender.

Yield: 4 servings.

Calcium: 153 mg. per serving.

Calories: 237 per serving.

SHRIMP, RICE, AND PEAS

1 tablespoon margarine
1 small onion, chopped
1 1/4 pounds shrimp, shelled and deveined
1/2 teaspoon dried dillweed
10 1/2-ounce can cream of mushroom soup
10 1/2 ounces skim milk
1 1/2 cups quick-cooking rice
10-ounce package frozen peas
4 ounces Cheddar cheese, grated

1. Melt margarine in a skillet and sauté onion till limp. Add shrimp and dill; brown slightly.

2. Add soup, milk, rice, and peas. Cover and simmer 10–12 minutes, or until liquid is absorbed and rice is cooked.

3. Stir in cheese and cook until melted.

Yield: 6 servings.

Calcium: 296 mg. per serving.

Calories: 386 per serving.

FISH ROLLED WITH SPINACH STUFFING

- 1 tablespoon oil
- 1 medium onion, chopped
- 2 garlic cloves, minced
- 10-ounce package frozen chopped spinach, thawed and drained
- 1 tablespoon lemon juice
- ½ cup part-skim ricotta cheese
- 1 egg, beaten
- 2 tablespoons flour
- Salt and pepper to taste
- 1 teaspoon dried dillweed
- ½ teaspoon ground allspice
- 2 pounds sole or flounder fillets
- ¼ cup dry white wine

1. Heat oil; sauté onion and garlic until softened.

2. Stir in spinach and cook until softened. Stir in lemon juice. Cool to room temperature.

3. Add cheese, egg, flour, and seasonings.

4. Spoon 2 tablespoons of mixture onto the center of each fillet. Roll up fillets and place seam side down in a 9x13-inch baking dish. Pour in wine.

5. Bake 20 minutes in a 375° oven, or until fish flakes when tested with a fork.

Yield: 8 servings.

Calcium: 120 mg. per serving.

Calories: 105 per serving.

PIZZA FILLETS FLORENTINE

- 10 ounces frozen chopped spinach, thawed and drained
- 1 pound fish fillets
- Salt and pepper
- 8-ounce can tomato or pizza sauce
- 2 garlic cloves, minced
- 1 teaspoon oregano
- ½ teaspoon dried basil
- ½ pound part-skim mozzarella cheese, shredded

1. Distribute spinach evenly in a lightly greased baking dish.

2. Place fillets over spinach. Sprinkle with salt and pepper.

3. Simmer tomato sauce with garlic and spices for 5 minutes. Pour over fish.

4. Top with mozzarella.

5. Bake, uncovered, in 375° oven for 25–30 minutes, or until fish flakes and cheese begins to brown.

Yield: 4 servings.

Calcium: 533 mg. per serving.

Calories: 228 per serving.

BAKED FISH AND POTATOES

- 1 medium onion, diced
- ½ green pepper, diced
- 1 tablespoon oil
- ⅛ teaspoon ground pepper
- 15½-ounce can salmon, drained
- 1 pound new potatoes, boiled
- ¼ cup skim milk
- 2 eggs, beaten
- Paprika

1. Sauté onion and pepper in oil until softened. Season with pepper. Remove from heat.

2. Mash salmon with potatoes. Stir in milk, eggs, and then the sautéed vegetables.

3. Transfer to greased pie plate or small baking dish. Sprinkle with paprika.

4. Bake in a 400° oven for 30 minutes until lightly browned.

Yield: 6 servings.

Calcium: 252 mg. per serving.

Calories: 345 per serving.

SCALLOPS WITH ALMOND COATING

- 1/3 cup flour
- 1 teaspoon dried tarragon or thyme, crumbled
- Salt and pepper
- 1 egg
- 2 tablespoons skim milk
- 2/3 cup cracker crumbs or matzoh meal
- 1/2 cup ground almonds
- 24 bay scallops or 12 halved sea scallops (approximately 1 pound)
- Oil to sauté fish*

1. On a large plate, mix flour, tarragon, salt, and pepper.

2. In a large bowl, beat the egg with the milk.

3. On another plate, mix crumbs with almonds.

4. Dredge scallops in flour, dip in egg-milk mixture, then coat with crumbs.

5. Heat oil in a skillet; sauté scallops until golden. Serve with lemon wedges.

Yield: 4 servings.

Calcium: 110 mg. per serving.

Calories: 257 per serving.

* Not included in calorie and calcium calculations.

SEVICHE WITH VEGETABLES

½ pound whole small mushrooms
1 pound bay scallops or halved sea scallops
2 cups fresh lime juice
¼ cup oil
½ cup white wine
3 garlic cloves, minced
2 canned green chilies, diced
1 teaspoon oregano
2 fresh tomatoes, diced
1 large avocado, cubed
½ cup chopped scallions
2 tablespoons chopped cilantro or parsley

1. Clean and dry mushrooms; remove stems. Place scallops and mushroom caps in a bowl.

2. Combine lime juice, oil, wine, garlic, chilies, and oregano; pour over scallops and mushrooms. Refrigerate at least 8 hours; turning from time to time.

3. Two hours before serving, mix in tomatoes, avocado, scallions, and cilantro; refrigerate until ready to serve.

Yield: 8 servings.

Calcium: 126 mg. per serving.

Calories: 205 per serving.

SEAFOOD CREPES

2 tablespoons margarine
2 tablespoons flour
1 cup fish stock or bottled clam juice
½ cup skim milk
¾ cup cream of mushroom soup, undiluted
4 ounces grated Cheddar or Monterey Jack cheese
Pepper to taste
½ pound cooked crab meat, flaked
8 crepes (recipe follows)

1. Melt magarine; stir in flour and cook for 2 minutes.

2. Add stock, milk, and soup; allow to thicken. Stir in cheese and pepper; allow cheese to melt.

3. Remove 1 cup of completed sauce. Mix remainder with crab meat and use as filling for crepes.

4. Roll crepes and place seam side down in a greased baking dish.

5. Spoon reserved sauce over crepes. Bake uncovered in a 400° oven for 15–20 minutes until lightly browned.

Yield: 8 crepes.

Calcium: 203 mg. per crepe (includes crepe).

Calories: 216 per crepe (includes crepe).

CREPES

1½ cups flour
2 eggs
1 tablespoon margarine, melted
2 cups skim milk
1 teaspoon sugar
½ teaspoon salt
Oil

1. Mix the flour with the eggs and melted margarine until smooth.

2. Add the milk, sugar, and salt; mix until completely blended.

3. Heat a small skillet (8 inches or less). Brush with a little oil and pour in a little batter, tilting pan so that batter forms a thin layer in bottom. Immediately pour excess batter back into the bowl of batter. Cook only till lightly browned. Turn and cook on other side.

4. Fill with whatever filling desired.

Yield: 12 crepes.

Calcium: 59 mg. per crepe.

Calories: 95 per crepe.

Note: Crepes should be thin and crisp. Do not use too much batter in pan.

HOT POTS

- 6 cups chicken or vegetable stock
- 24 raw shrimp, peeled and deveined
- 12 raw clams (littlenecks or cherrystones), scrubbed
- 12 medium mushrooms
- ½ pound Chinese cabbage, washed and separated
- ½ pound spinach leaves, washed
- ½ pound firm tofu, cut into bite-size pieces
- ½ pound broccoli, cut into florets
- 2 tablespoons soy sauce
- 2 tablespoons dry sherry
- 1 tablespoon finely chopped fresh ginger
- 3 scallions, chopped
- 3 cups hot cooked rice

An Oriental hot pot is similar to a fondue, but instead of cheese, chocolate or oil, the cooking liquid is a broth. You will need one skewer per person, maybe two if you want to be able to eat more quickly. Each person skewers pieces of vegetable and fish and places in boiling broth until food is cooked. Foods are eaten plain or dipped in soy sauce, plum sauce, or other sauces available in Chinese markets. At the end, the broth becomes a soup by seasoning with soy sauce, sherry, ginger, and scallions. Cooked rice is added to the broth before serving as soup.

Yield: 6 servings.

Calcium: 219 mg. per serving.

Calories: 272 per serving.

CINNAMON-NUT BLINTZES

- 1 pound farmer cheese
- ½ cup chopped figs or raisins
- ½ cup brown sugar
- 2 eggs, beaten
- 1 teaspoon cinnamon
- 12 crepes (see Index)
- 1 tablespoon margarine

1. Combine cheese, figs, sugar, eggs, and cinnamon.

2. Place 2 tablespoons of filling in center of each crepe. Fold ends in, then roll.

3. Melt margarine in a hot skillet; place crepes in skillet. Brown on one side, then turn and brown on second side.

4. Serve with yogurt and preserves or apple sauce, if desired.

Yield: 12 blintzes.

Calcium: 281 mg. per blintz.

Calories: 273 per blintz.

MUSHROOM AND RICOTTA CHEESE PIE

1 onion, chopped
1 tablespoon margarine
½ pound fresh mushrooms, sliced
½ teaspoon dried basil
¼ teaspoon ground pepper
2 whole eggs
2 egg whites, beaten
3–4 tablespoons flour
16 ounces ricotta or low-fat cottage cheese
Generous sprinkling of nutmeg
9-inch pie shell, unbaked
1 cup low-fat plain yogurt
Generous sprinkling of paprika

1. Sauté onion in margarine until limp. Add mushrooms and cook until softened. Add basil and pepper. Set aside.

2. Beat whole eggs with egg whites. Stir in flour. Stir in ricotta and nutmeg. Stir in mushroom mixture.

3. Pour into pie shell. Top with yogurt and sprinkle with paprika.

4. Bake in a 375° oven for 40–45 minutes until set.

Yield: 6 servings.

Calcium: 147 mg. per serving.

Calories: 332 per serving.

MUSHROOM AND CHARD CHEESE SOUFFLÉ

- 3 tablespoons margarine
- ½ onion, diced
- 1½ cups sliced mushrooms
- 3 cups chopped fresh Swiss chard or spinach leaves (about 1¼ pounds)
- 3 tablespoons flour
- 3 eggs, separated
- ¼ teaspoon salt
- ¼ teaspoon paprika
- ½ cup crumbled feta or blue cheese

1. Melt 1 tablespoon of the margarine. Sauté onion, mushrooms, and then chard until softened.

2. Stir in and melt remaining margarine. Add flour and cook, stirring constantly, for 2 minutes. Remove from heat and cool.

3. Stir in egg yolks, salt, and paprika; stir in cheese.

4. Preheat oven to 375°. Beat egg whites until stiff peaks form. Gently fold into vegetable mixture.

5. Pour into 2-quart casserole or soufflé dish and bake 35 minutes, or until eggs are set to desired consistency.

Yield: 4 servings.

Calcium: 235 mg. per serving.

Calories: 244 per serving.

BAKED MUSHROOMS AND CHEESE

- 12 large mushrooms
- 2 tablespoons margarine
- 1 cup White Sauce (see Index)
- 1 cup grated Cheddar cheese
- Dash of white pepper
- ⅓ cup seasoned bread crumbs
- Paprika

1. Wash mushrooms. Remove stems. Chop stems and sauté in margarine until tender.

2. Stir in White Sauce, cheese, and pepper. Cook until heated.

3. Arrange mushroom caps, cut side up, in a baking dish. Pour sauce over mushrooms.

4. Sprinkle bread crumbs over each mushroom. Sprinkle with paprika.

5. Bake in 350° oven for 25 minutes.

Yield: 4 servings.

Calcium: 303 mg. per serving.

Calories: 339 per serving.

CHEESE FONDUE

1 garlic clove
¾ cup dry white wine
8 ounces Swiss cheese, grated
1 tablespoon flour
⅛ teaspoon white pepper
2 tablespoons kirsch
4 cups cubed French or Italian bread

1. Rub the sides of a fondue pot with garlic. Pour in wine and heat until bubbly.

2. Toss the cheese with the flour and stir into the wine until melted.

3. Stir in the pepper and kirsch.

4. Dip bread cubes into sauce.

Yield: 4 servings.

Calcium: 541 mg. per serving (excluding bread).

Calories: 273 per serving (excluding bread).

Note: Add 9 mg. calcium and 58 calories for each 1-ounce slice of bread.

KALE AND TRIPLE CHEESE PIE

1 tablespoon margarine
½ cup chopped onion
3 garlic cloves, minced
1¼ pounds of kale
¼ cup chopped fresh parsley
3 whole eggs
2 egg whites
1 pound low-fat cottage cheese
½ pound feta cheese, crumbled
¼ cup grated Parmesan cheese
⅛ teaspoon ground pepper
2 8-inch pie shells, unbaked

1. Melt margarine; sauté onion and garlic till limp.

2. Wash kale and pat dry. Remove stems and coarsely chop leaves. Add to onion mixture and cook until softened, adding water if necessary.

3. Add parsley and remove from heat.

4. Beat whole eggs with egg whites. Stir in cheeses and pepper. Stir in vegetable mixture.

5. Pour into pie shells. Bake 30 minutes at 375° until browned on top.

Yield: 12 or 16 servings (6 or 8 servings per pie).

Calcium: 222 mg. per serving if 16 servings; 250 if 12.

Calories: 262 per serving if 16 servings; 299 if 12.

STUFFED EGGPLANT WITH CHEESE

- 1 onion, chopped
- 1 green pepper, cut into strips
- 2 garlic cloves, minced
- 1 tablespoon vegetable oil
- 2 medium eggplants
- 15-ounce can kidney beans, drained
- 1 cup tomato sauce
- 1 teaspoon chili powder
- 1 cup shredded mozzarella cheese
- 1/4 cup grated Parmesan cheese

1. Sauté onion, peppers, and garlic in oil until softened.

2. Cut eggplants in half and scoop out pulp, leaving a 1/4-inch shell. Dice pulp and add to cooked vegetables; cook and stir for 5 minutes, adding drops of water to prevent sticking.

3. Add beans, tomato sauce, and chili powder; simmer 10 minutes until eggplant is tender.

4. Preheat oven to 375°. Spoon the mixture into the reserved eggplant shells. Sprinkle with cheeses.

5. Pour a little water into the bottom of a baking dish. Place shells in dish. Bake 30 minutes, or until shells are tender when pierced with a fork.

Yield: 4 servings.

Calcium: 363 mg. per serving.

Calories: 326 per serving.

PUFFED CHEESE, TOMATOES, AND OLIVES BAKE

2 garlic cloves, minced
1 tablespoon vegetable oil
1½ cups chopped fresh tomatoes
⅓ cup sliced black olives
½ teaspoon dried basil
4 eggs, beaten
⅛ teaspoon salt
Dash of pepper
1 cup skim milk
¾ cup flour
1 cup shredded part-skim mozzarella or Monterey Jack cheese

1. Sauté garlic in oil until softened. Stew tomatoes in oil until liquid evaporates.

2. Remove from heat, stir in olives and basil. Spoon into greased 9-inch pie pan. Preheat oven to 400°.

3. In a bowl, combine eggs, salt, pepper, and half the milk. Stir in flour, beating until there are no lumps.

4. Add remaining milk, then cheese. Pour over tomato mixture. Bake 25 minutes, or until puffed and golden.

Yield: 4 servings.

Calcium: 360 mg. per serving.

Calories: 350 per serving.

TAMALE PIE

CRUST:

1 cup cold water
1 cup skim milk
1 cup corn meal
½ teaspoon salt
½ teaspoon chili powder

Filling:

- 1 tomato, diced
- 2 zucchini (approximately ½ pound), sliced
- 8-ounce can kidney beans, drained
- 8 ounces Monterey Jack or Cheddar cheese, grated
- ½ cup sliced ripe olives

Garnish:

- 1 large avocado, sliced

Hot Sauce (optional):

- 1 cup tomato sauce
- 2 garlic cloves, crushed
- ¾ teaspoon oregano
- ½ teaspoon cumin
- ⅛ teaspoon cayenne pepper

1. Pour all the crust ingredients into a saucepan. Simmer and stir over moderate heat until stiff and thick; cool.

2. Press mixture into bottom and sides of a greased small casserole dish.

3. Spoon vegetables and beans into crust. Top with cheese and olives. Bake at 375° for 40 minutes.

4. Garnish with avocado slices.

5. If desired, tamale pie can be served with a hot sauce prepared by simmering all the sauce ingredients for 15 minutes.

Yield: 6 servings.

Calcium: 449 mg. per serving.

Calories: 412 per serving (includes hot sauce).

BEAN BURRITOS

- 1 tablespoon oil
- 1 large onion, chopped
- 1 cup diced tomatoes
- 1 teaspoon salt
- 1/2 teaspoon ground pepper
- 1/4 teaspoon cumin
- 1/8 teaspoon hot pepper sauce (e.g., Tabasco)
- 15 1/2-ounce can vegetarian baked beans
- 12 large flour tortillas, uncooked
- 3/4 cup grated Cheddar cheese
- 1/2 cup sliced olives
- 1/2 cup chopped scallions

1. Sauté onion in oil until limp. Add tomatoes, salt, pepper, cumin, and hot pepper sauce and simmer for 10 minutes. Stir in beans and simmer 5 minutes longer.

2. Spoon 2 tablespoons mixture into each tortilla. Roll and place seam side down in a baking dish.

3. Sprinkle with cheese, olives, and scallions. Bake for 15 minutes until cheese is melted. Serve with sour cream, if desired.

Yield: 12 burritos.

Calcium: 152 mg. per serving.

Calories: 164 per serving.

ASPARAGUS ENCHILADAS WITH NUTTY WHITE SAUCE

- 1 tablespoon margarine
- 1 tablespoon flour
- 1/4 cup ground almonds
- 1/3 cup low-fat plain yogurt
- 2/3 cup skim milk
- Dash of cayenne
- 1 large (15-ounce) can asparagus tips, drained
- 4 flour tortillas, uncooked
- 1/4 cup shredded part-skim mozzarella, or other mild-flavored white cheese

1. Prepare sauce: Melt margarine, add flour, and cook 2 minutes, stirring constantly. Stir in almonds and cook 30 seconds. Add yogurt, milk, and cayenne. Cook until thickened. Remove from heat.

2. Preheat oven to 350°. Spread 1 tablespoon sauce on each tortilla. Arrange asparagus on center of tortillas. Roll and place seam side down in a baking dish.

3. Spoon remaining sauce over tortillas. Sprinkle with cheese. Bake until bubbly and melted, about 15 minutes.

Yield: 2 servings.

Calcium: 464 mg. per serving.

Calories: 417 per serving.

CHILI RELLENO CASSEROLE

- 3 7-ounce cans whole green chilies
- 1 pound Monterey Jack cheese, grated
- 1 pound Cheddar cheese, grated
- 3 eggs
- 3 tablespoons flour
- Small can (approximately ¾ cup) evaporated milk
- 15-ounce can tomato sauce

1. Wash chilies, remove seeds, and pat dry.

2. In a greased 9x13-inch baking dish, layer half the chilies, then half the cheeses. Repeat, reserving ¼ cup of each cheese for topping.

3. Beat eggs, stir in flour, then milk. Pour into casserole.

4. Bake in a 350° oven for 30 minutes. Spread tomato sauce over the top. Sprinkle with reserved cheese and bake 15 minutes longer. Cut into pieces.

Yield: 12 servings.

Calcium: 615 mg. per serving.

Calories: 348 per serving.

CHEESE ENCHILADAS

1 cup chopped onion
1 tablespoon vegetable oil
15-ounce can tomato sauce
1 teaspoon cumin
Dash of black pepper and cayenne pepper
12 corn tortillas
2 cups grated Cheddar or Monterey Jack cheese
6 scallions, chopped
7-ounce can green chilies, chopped
3/4 cup sliced black olives

1. Sauté onion in oil until softened. Add tomato sauce, cumin, and pepper; simmer until reduced by one third. Set aside.

2. Wrap tortillas in foil and heat in 350° oven for 5 minutes, or until softened.

3. To prepare enchiladas, spoon small amount of sauce on each tortilla; sprinkle with 2 tablespoons cheese, 2 teaspoons scallions, 1 teaspoon chilies, and 1 tablespoon olives. Roll tortilla and place seam side down in a greased baking dish.

4. Thin the remaining sauce with a little water or tomato juice. Pour over tortillas and top with remaining cheese.

5. Bake in 350° oven for 20 minutes, or until hot.

Yield: 6 servings.

Calcium: 448 mg. per serving.

Calories: 386 per serving.

SPINACH PIE

1 tablespoon margarine
1 small onion, minced
4 bunches spinach (about 2 1/2 pounds), cleaned, chopped, steamed, and drained
4 eggs, beaten
4 cups scalded skim milk
1/2 pound Cheddar cheese, grated
2 8-inch pie shells, unbaked

1. In a large skillet, melt the margarine and sauté onion. Stir in spinach and mix with onion.

2. Spoon into a large bowl. Stir in eggs, then milk, then cheese. Preheat oven to 425°.

3. Pour mixture into pie shells. Bake for 10 minutes, reduce heat to 350°, and bake until knife inserted in center comes out clean.

Yield: 2 pies of 8 slices each.

Calcium: 260 mg. per slice.

Calories: 211 per slice.

RATATOUILLE CREPE

1 tablespoon oil
2 small onions, minced
½ pound mushrooms, minced
4 garlic cloves, minced
¾ pound zucchini, minced
1 small eggplant (approximately ½ pound), minced but unpeeled
15½-ounce can tomato sauce
1 teaspoon basil
½ teaspoon salt
¼ teaspoon pepper
6 ounces mozzarella cheese, grated
4 tablespoons grated Parmesan cheese
12 crepes (see Index)

1. Heat the oil, add all the vegetables and stir-fry until tender, adding bits of water to prevent sticking.

2. Add tomato sauce, basil, salt, and pepper; simmer 15 minutes. Remove from heat.

3. Stir in the mozzarella and 2 tablespoons of the Parmesan cheese.

4. Set aside one third of the filling. Divide remaining filling evenly among the crepes, about 1½ tablespoons per crepe. Roll crepes

(*recipe continues*)

and place in lightly greased baking dish. Spoon remaining filling on top, sprinkle with remaining Parmesan cheese and bake in a 400° oven for 15 minutes.

Yield: 12 crepes.

Calcium: 195 mg. per crepe.

Calories: 197 per crepe.

CURRIED CAULIFLOWER CREPES

1 head cauliflower, broken into florets
1 medium onion, chopped
2 garlic cloves, minced
2 tablespoons oil
1 teaspoon curry powder
½ teaspoon cumin
½ teaspoon turmeric
¼ teaspoon cayenne
Salt to taste
2 tomatoes, chopped
½ cup chopped figs
2 cups low-fat plain yogurt
16 crepes (see Index)
½ cup sliced almonds, lightly toasted

1. Steam cauliflower till tender. When cool, chop and reserve.

2. Sauté onion and garlic in oil until tender. Stir in spices and salt and cook 1 minute.

3. Add tomatoes, figs, and 1 cup of the yogurt. Gently stir in cauliflower.

4. Spoon 2 tablespoons filling on each crepe. Roll crepes and place in a greased baking dish. Pour remaining yogurt over crepes; sprinkle with almonds.

5. Bake 20 minutes in a 375° oven.

Yield: 16 crepes.

Calcium: 133 mg. per crepe.

Calories: 142 per crepe.

STUFFED ZUCCHINI

- 1 medium onion, diced
- 3 garlic cloves, minced
- 1 tablespoon sesame or peanut oil
- 2 medium zucchini
- 1 pound soft tofu, drained and mashed
- 1/4 cup soy sauce
- 2 teaspoons minced fresh ginger, or 1/2 teaspoon powdered ginger

1. Sauté onion and garlic in oil until onion is limp.

2. Cut zucchini in half lengthwise. Scoop out pulp, leaving a 1/4-inch shell. Chop pulp.

3. Add zucchini pulp and tofu to onion and cook until softened, adding more oil if needed to prevent sticking.

4. Stir in soy sauce and ginger.

5. Fill zucchini shells with mixture.

6. Place in baking dish, pour a little hot water in the bottom of the dish, and cover.

7. Bake in 350° oven until shells are tender (30–40 minutes).

Yield: 4 servings.

Calcium: 173 mg. per serving.

Calories: 148 per serving.

Pasta and Rice

Pasta, long neglected by Americans who mistakenly thought it was fattening, is being rediscovered as a nutritious way to control appetites and, in fact, maintain ideal weight. The following recipes show how pasta and rice can be used as the foundation for high-calcium entrées or side dishes. If you are trying to lose weight, calories can be reduced by eliminating or reducing nuts and seeds, or by reducing the amount of cheese.

NOODLE KUGEL

- 1 pound broad noodles, cooked
- 1 pound low-fat cottage cheese
- 4 eggs, beaten
- 1 cup low-fat plain yogurt
- 1 cup golden raisins or chopped Calimyrna figs
- 3/4 cup brown sugar
- 4 tablespoons margarine, melted
- 1 1/2 teaspoons vanilla extract
- 1 tablespoon margarine
- Cinnamon

1. Combine noodles, cottage cheese, eggs, yogurt, raisins or figs, brown sugar, melted margarine, and vanilla. Pour into a greased casserole dish. Refrigerate overnight.

2. Dot with 1 tablespoon margarine and sprinkle with cinnamon.

3. Bake in a 350° oven for 1 1/2 hours, or until golden brown on top.

Yield: 8 servings.

Calcium: 144 mg. per serving.

Calories: 498 per serving.

SUNFLOWER SEED PESTO OVER GREEN NOODLES

- 8 ounces spinach noodles
- 1/2 cup frozen peas
- 1 1/2 cups romaine leaves, cleaned
- 1/3 cup shelled sunflower seeds
- 1/4 cup fresh lemon juice
- 1/4 cup grated Parmesan cheese
- 1/2 teaspoon salt
- 1/8 teaspoon ground pepper
- 6 fresh basil leaves, or 1 teaspoon dried basil
- 3 scallions, chopped
- 3 garlic cloves, minced
- 1/3 cup vegetable oil

1. Bring 6 cups of water to a boil. Add noodles and cook until tender, adding peas 3 minutes before completed; drain.

2. Meanwhile, combine remaining ingredients in a blender or food processor and whirl until finely minced, but not puréed.

3. Stir 1/2 cup of the sauce into noodles and peas. Serve immediately.

Yield: 4 servings.

Calcium: 162 mg. per serving.

Calories: 493 per serving.

Note: The remaining pesto sauce can be added to soups, used as a spread on sandwiches, or used as a dressing if thinned with fresh lemon juice.

PASTA AND CHEESE

8 ounces spinach pasta
2 tablespoons margarine
2 tablespoons flour
1 1/3 cups skim milk
1 cup grated Cheddar cheese
1/4 teaspoon white pepper
1/4 cup chopped pimientos

1. Cook pasta according to package directions; drain.

2. Melt margarine in a small saucepan. Stir in flour and cook 2 minutes.

3. Stir in milk and cook until thickened.

4. Stir in cheese, pepper, and pimientos and heat until cheese is melted.

5. Add pasta and serve, or pour into a greased baking dish and bake in a 350° oven for 20 minutes.

Yield: 5 servings.

Calcium: 280 mg. per serving.

Calories: 338 per serving.

SPAGHETTI WITH CHEESE SAUCE

- 1 cup shredded part-skim mozzarella, garlic, or pepper cheese
- ½ cup grated Parmesan cheese
- ¼ cup skim milk
- ¼ cup finely minced fresh parsley
- 2 eggs, beaten
- Ground pepper to taste
- ½ cup sliced toasted almonds
- 8 ounces spaghetti, cooked and drained

Combine and heat cheeses, milk, parsley, eggs, pepper, and almonds. Toss with cooked spaghetti.

Yield: 6 servings.

Calcium: 339 mg. per serving.

Calories: 327 per serving.

BAKED MACARONI AND CHEESE WITH SALMON

- 3 tablespoons margarine
- 1 small onion, diced
- 1 tablespoon flour
- ¼ teaspoon dry mustard
- Generous sprinkling pepper
- 1½ cups skim milk
- 1 tablespoon minced fresh parsley
- 1 cup shredded Cheddar cheese
- 2 cups elbow macaroni, cooked and drained
- 14 ounces canned salmon, drained and broken into chunks
- 1 cup frozen peas, cooked and drained
- ¾ cups fresh bread crumbs

1. Melt 1 tablespoon of the margarine and sauté onion until limp.

Add flour, mustard, and pepper; cook for 2 minutes.

2. Slowly stir in milk and parsley; cook until smooth and thickened. Remove from heat and stir in cheese.

3. In a bowl, combine cooked macaroni with salmon and peas.

4. Preheat oven to 350°. Spoon macaroni mixture into greased casserole. Pour cheese sauce over macaroni.

5. Melt remaining 2 tablespoons margarine. Stir and sauté bread crumbs until well coated. Sprinkle crumbs over top of macaroni mixture.

6. Bake 25 minutes until bubbly and crumbs are golden.

Yield: 6 servings.

Calcium: 370 mg. per serving.

Calories: 340 per serving.

PASTA E FAGIOLI WITH TOMATO SAUCE

1 cup pasta twists or elbow macaroni
1 cup cooked kidney beans
1½ cups tomato sauce
⅛ teaspoon hot pepper sauce (e.g., Tabasco)
½ cup diced zucchini
½ cup shredded part-skim mozzarella cheese

1. Boil pasta until tender. Drain.

2. Mix beans with tomato sauce, hot pepper sauce, and zucchini; simmer 15 minutes. Stir in cooked pasta, then cheese. Cook until heated and cheese is melted.

Yield: 2 servings.

Calcium: 305 mg. per serving.

Calories: 467 per serving.

LINGUINE WITH MUSSEL AND TOMATO SAUCE

16-ounce can stewed tomatoes
1 small onion, chopped
2 tablespoons tomato paste
1 teaspoon dried fennel seeds
1/4 cup dry white wine
2 pounds mussels, cleaned and steamed
8 ounces linguine or spaghetti, cooked
1/4 cup finely chopped fresh parsley

1. Simmer the tomatoes and their liquid with the onion, tomato paste, and fennel until sauce is thickened, approximately 20 minutes.

2. Pour in the wine and cook 3 minutes longer.

3. Remove steamed mussels from their shells (save the broth for another use). Chop the mussels coarsely and add to sauce. Heat briefly.

4. Serve over cooked pasta. Sprinkle with parsley.

Yield: 4 servings.

Calcium: 255 mg. per serving.

Calories: 469 per serving.

TOMATO SAUCE LASAGNE

- 1 small onion, chopped
- 3 garlic cloves, minced
- 1 tablespoon vegetable oil
- 1-pound can tomatoes, coarsely chopped
- 3 ounces tomato paste
- ½ cup water
- 2 tablespoons chopped fresh basil, or 1 teaspoon dried
- 1 teaspoon dried oregano
- 1 pound part-skim ricotta cheese or low-fat cottage cheese
- ½ pound part-skim mozzarella cheese, shredded
- 1 egg, beaten
- 2 tablespoons minced fresh parsley
- ¼ teaspoon ground pepper
- ¼ cup grated Parmesan cheese
- 8 ounces lasagne noodles, cooked and separated

1. In a large saucepan, sauté onion and garlic in oil until limp. Add tomatoes with their liquid, tomato paste, water, basil, and oregano. Simmer for 30–40 minutes, or until thickened but still slightly liquid.

2. Mix ricotta, mozzarella, egg, parsley, and pepper.

3. Spoon a little sauce into the bottom of an 8-inch square baking pan. Make three layers of noodles, sauce, and cheese mixture.

4. Spoon some sauce over the top cheese layer and sprinkle with Parmesan.

5. Bake in a 375° oven for 45 minutes. Let sit 10 minutes before serving.

Yield: 6 servings.

Calcium: 489 mg. per serving.

Calories: 450 per serving.

LASAGNE WITH PEAS AND MUSHROOMS

9 lasagne noodles, cooked and drained
1 pound cottage cheese (1% fat)
1/3 cup skim milk
1/2 teaspoon garlic powder
1/4 teaspoon dried oregano
1/4 teaspoon dried basil
1/4 teaspoon dried thyme
1 tablespoon margarine
1 medium onion, chopped
1 pound mushrooms, sliced
10 ounces frozen peas, thawed
1 cup grated Parmesan cheese

1. Mix cottage cheese with milk, garlic powder, and herbs until creamy; set aside.

2. Melt margarine in a skillet; sauté onion, then mushrooms, till tender.

3. In a 9x13-inch-pan, layer as follows: 3 noodles, 1/3 of cheese mixture, 1/3 of onion-mushroom mixture, 1/3 of peas.

4. Repeat this layering pattern twice. Top with grated Parmesan. Bake at 350° for 1 hour.

Yield: 6 servings.

Calcium: 464 mg. per serving.

Calories: 373 per serving.

RICE, FRUIT, AND NUT SALAD

1 cup converted rice
1 teaspoon cinnamon
1 bay leaf
1/8 teaspoon salt
1 cup shredded carrots
1/2 cup sliced almonds
1/2 cup chopped dried figs

1. Cook rice until tender in 2½ cups water to which cinnamon, bay leaf, and salt have been added; remove from heat and cool to room temperature.

2. Stir in carrots, almonds, and figs.

3. Toss with Lime and Sesame Seed Dressing (see Index), but do not add molasses or maple syrup to dressing.

Yield: 8 servings.

Calcium: 31 mg. per serving.

Calories: 139 per serving (excluding dressing).

VEGETABLE PILAF

1 medium onion, chopped
1 celery rib, sliced
1 tablespoon oil
1½ cups chopped kale or spinach
1 cup sliced asparagus or green beans
½ cup chopped tomatoes
1 medium rutabaga, peeled and cubed
2½ cups water
1 cup converted rice
1 teaspoon thyme
Salt, if desired

1. Sauté onion and celery in oil in large saucepan until onion is limp.

2. Add vegetables and 1 cup of the water; cover and simmer 10 minutes, stirring occasionally.

3. Add rice, remaining 1½ cups water, thyme, and salt. Cover and simmer until rice is soft but not mushy.

Yield: 8 servings.

Calcium: 41 mg. per serving.

Calories: 129 per serving.

RICE WITH SWEET POTATOES AND GARBANZO BEANS

- 3/4 cup chopped onion
- 1 tablespoon oil
- 1/2 pound sweet potatoes, peeled and cubed
- 2 1/3 cups water
- 1 cup orange juice
- 1 1/2 cups converted rice
- Salt to taste
- 1 teaspoon dried thyme
- 1/2 cup frozen peas
- 1/2 cup cooked garbanzo beans

1. Sauté onion in oil until limp.

2. Add sweet potatoes and a bit of the water to keep potatoes from sticking. Cook until softened slightly.

3. Add remaining water, juice, rice, salt, and thyme and cook until almost all water is absorbed.

4. Add peas and garbanzo beans; cook until heated throughout.

Yield: 8 servings.

Calcium: 59 mg. per serving.

Calories: 253 per serving.

MEXICAN RICE AND BEANS

- 1 tablespoon vegetable oil
- 1 medium onion, chopped
- 1 medium green pepper, chopped
- 2 cups tomato sauce
- 1 teaspoon chili powder
- 1 teaspoon dried oregano
- 1 1/2 cups boiling water
- 1 cup long-grain rice
- 2 cups cooked or canned kidney beans, drained
- 1 1/2 cups shredded cheese (Cheddar, Monterey Jack, American, or jalapeño) (optional)

1. Heat oil in a large pot. Sauté onion and pepper until lightly browned.

2. Add tomato sauce, chili powder, and oregano. Simmer 15 minutes, stirring occasionally. Add water and bring to a strong simmer.

3. Stir in rice. Cover and simmer for 20 minutes, or until rice is tender.

4. Stir in kidney beans and heat thoroughly. If desired, pour mixture into greased baking dish and sprinkle with cheese. Bake in 375° oven until cheese is bubbly.

Yield: 6 servings.

Calcium: 276 mg. per serving (includes cheese).

Calories: 285 per serving (includes cheese).

Legumes, Potatoes, and Corn

These foods are starches, or complex carbohydrates. Like pasta, they are often shunned as fattening, but in reality they have fewer calories ounce for ounce than meat, and they provide many important nutrients. The following recipes can be used as entrées, salads, or side dishes, and all provide good amounts of calcium.

LENTIL PILAF

1 tablespoon oil
1 medium onion, chopped
3 cups water
1 tablespoon vegetable bouillon powder
½ cup lentils
½ cup converted rice
2 cups chopped fresh spinach leaves, or 5 ounces frozen chopped spinach, thawed and drained
2 carrots, cut into 1-inch matchsticks
2 parsnips, cut into 1-inch matchsticks

1. Heat oil in a saucepan, add onion, and cook until limp.

2. Add water and bouillon; bring to a boil.

3. Add lentils, cover, and simmer for 30 minutes.

4. Add rice, spinach, carrots, and parsnips. Cook until tender (approximately 20 minutes), adding more water if needed.

Yield: 6 servings.

Calcium: 89 mg. per serving.

Calories: 206 per serving.

SPICY LENTILS

1½ cups lentils
4½ cups salted water
½ onion, chopped
2 garlic cloves, minced
1 tablespoon sesame or vegetable oil
2 tablespoons vinegar
1 teaspoon crushed red pepper
Dash of black pepper
Juice of 1 lime or ½ lemon
Romaine lettuce leaves
Garnishes: pimientos and scallions*

1. Cook lentils in the water until tender (approximately 1 hour). Drain if necessary; pour into a large bowl.

2. Sauté onion and garlic in oil until lightly browned. Remove from heat and add to lentils.

3. Stir in vinegar, crushed red pepper, and black pepper. Sprinkle with juice. Chill.

4. Serve on romaine lettuce. Garnish with pimientos and scallions.

Yield: 6 servings.

Calcium: 26 mg. per serving.

Calories: 201 per serving.

* Not included in calorie and calcium calculations.

BARBECUED BEANS WITH VEGETABLES

1½ cups cooked pinto beans
1 cup grated carrots
1 cup chopped broccoli
¼ cup barbecue sauce
¼ cup ground almonds
¼ cup water
½ cup grated Cheddar cheese

1. Mash half the beans.

2. Combine mashed and whole beans with remaining ingredients except the cheese.

3. Pour into a greased casserole. Sprinkle with cheese and bake 30 minutes at 350°.

Yield: 4 servings.

Calcium: 272 mg. per serving.

Calories: 398 per serving.

BAKED BEANS

1 pound navy beans
7 cups water
1 cup tomato sauce
Scant 1/4 cup maple syrup or molasses
2 tablespoons Worcestershire sauce
1 tablespoon prepared mustard
1/2 teaspoon salt
1/4 teaspoon ground pepper

1. Rinse beans in cold water. In a large pot, bring beans and the 7 cups water to a boil. Boil 2 minutes, reduce heat, and simmer approximately 2 1/2 hours, or until beans are tender; drain.
2. In a medium-size baking dish, combine beans with remaining ingredients. Cover and bake at 350° for 1 hour. Uncover and bake 20 minutes longer.

Yield: 8 servings.

Calcium: 103 mg. per serving.

Calories: 234 per serving.

HOT RED AND WHITE BEAN SALAD

1 cup red beans
1 cup white or lima beans
2 garlic cloves, minced
3 onions, thinly sliced
3 teaspoons vegetable bouillon powder, or 3 cubes vegetable bouillon
1/4 cup cider vinegar
3 tablespoons molasses
3 tablespoons maple syrup
2 teaspoons salt
1/2 teaspoon ground pepper
Low-fat plain yogurt or sour cream*

1. Rinse beans, cover with water (2 inches above beans) and soak overnight. Drain and cover with fresh water 2 inches above beans.

2. Bring beans and water to a boil; reduce heat, cover, and simmer 30–40 minutes.

3. Drain off all but 3 cups water. Add garlic, onion, and bouillon.

4. In a bowl, mix remaining ingredients except yogurt. Add to beans. Pour entire mixture into an ovenproof dish and bake in a 300° oven for 3 hours. Reduce heat to 250°, uncover, and bake 1½ hours more, or until liquid is absorbed.

5. Serve hot with a dollop of yogurt or sour cream.

Yield: 6 servings.

Calcium: 134 mg. per serving.

Calories: 305 per serving.

* Not included in calorie and calcium calculations.

DILL AND CHEESE STUFFED POTATOES

4 baking potatoes
⅔ cup shredded Cheddar cheese
¼ cup skim milk
1 tablespoon margarine
1 teaspoon dried dillweed
⅛ teaspoon ground pepper

1. Bake potatoes. Cut a slice off top; scoop out pulp, leaving ¼-inch shell.

2. Mash pulp; add remaining ingredients.

3. Stuff into potatoes. Replace lids and heat 15 minutes at 350° until piping hot.

Yield: 4 servings.

Calcium: 174 mg. per serving.

Calories: 185 per serving.

LOW-CALORIE STUFFED POTATOES

4 baking potatoes
½ cup low-fat cottage cheese
½ cup low-fat plain yogurt
¼ cup chopped scallions, or 1 tablespoon chopped chives
Pepper to taste

1. Bake potatoes. Cut slice off top; scoop out pulp, leaving a ¼-inch shell.

2. Mash pulp; add remaining ingredients.

3. Stuff potatoes with mixture. Replace lids. Heat 15 minutes at 350° until piping hot.

Yield: 4 servings.

Calcium: 80 mg. per serving.

Calories: 117 per serving.

SPINACH AND PARMESAN POTATOES

4 baking potatoes
1 small onion, chopped
1 tablespoon vegetable oil
5 ounces frozen chopped spinach, thawed and drained
⅓ cup grated Parmesan cheese
1 teaspoon dried dillweed
Salt and pepper to taste

1. Bake potatoes. Cut slice off top; scoop out pulp, leaving ¼-inch shell. Mash the potato pulp.

2. Sauté onion in oil until tender. Add spinach and cook until liquid has evaporated. Add cheese, dillweed, seasoning, and mashed potatoes.

3. Spoon mixture into shells. Replace top of potato shells and heat 15 minutes at 350° until piping hot.

Yield: 4 servings.

Calcium: 150 mg. per serving.

Calories: 157 per serving.

SCALLOPED POTATOES WITH CHILIES AND CHEESE

3½ cups peeled and sliced potatoes
2 tablespoons margarine
3 tablespoons flour
1½ cups skim milk
1½ cups grated Cheddar cheese
7-ounce can mild green chilies, drained and diced
Salt, pepper, and onion powder to taste
Sprinkling of paprika

1. Boil potatoes until tender; drain well.

2. Prepare cheese sauce: Melt margarine, stir in flour, and cook 2 minutes. Gradually stir in milk, cheese, chilies, salt, pepper, and onion powder. Simmer until thickened.

3. In a lightly greased baking dish, place a layer of potatoes. Pour some cheese sauce over potatoes. Repeat with potatoes, then cheese sauce. Sprinkle with paprika.

4. Bake in a 350° oven for 50–60 minutes, or until golden brown.

Yield: 4 servings.

Calcium: 442 mg. per serving.

Calories: 291 per serving.

SPICED POTATO RAITA

1 tablespoon oil
1 teaspoon mustard seeds
3/4 cup low-fat plain yogurt
1/4 cup chopped scallions
1/4 teaspoon ground cumin
1/4 teaspoon salt
2 medium-size new potatoes, peeled, cooked, and cubed

1. Heat oil and sauté mustard seeds until they begin to pop.

2. Stir in yogurt, scallions, cumin, and salt.

3. Gently stir in potatoes. Chill. Serve as an accompaniment to spicy foods.

Yield: 4 servings.

Calcium: 134 mg. per serving.

Calories: 102 per serving.

SWEET POTATO PUDDING

4 cups cooked sweet potatoes
1 1/2 cups skim milk
1/2 cup sugar
1/2 cup diced figs
1/2 cup golden raisins
1/4 cup dark molasses
1/4 cup vegetable oil
4 eggs, beaten
Rind of 1 orange, grated
2 tablespoons flour
1/2 teaspoon ground cinnamon
1/4 teaspoon ground allspice
1 tablespoon margarine

1. Combine all ingredients except margarine. Pour into a large greased baking dish and dot with margarine.

2. Cover and bake in a 350° oven for 30 minutes. Uncover and bake 20 minutes longer. Serve hot.

Yield: 12 servings.

Calcium: 92 mg. per serving.

Calories: 269 per serving.

Note: Can be frozen after being cut into squares and cooled.

SWEET POTATO AND ALMOND PATTIES

4 cups peeled and cubed sweet potatoes
½ medium onion, grated
⅓ cup whole wheat flour
¼ cup ground almonds
½ teaspoon dried thyme
¼ teaspoon ground allspice
Salt and pepper to taste

1. Cook sweet potatoes and mash.

2. Combine sweet potatoes and all other ingredients and shape into 8 patties. Sauté both sides in just enough oil to cover bottom of skillet.

Yield: 8 patties.

Calcium: 46 mg. per patty.

Calories: 149 per patty.

TZIMMES

- ½ pound carrots
- 2 pounds sweet potatoes
- 6 ounces dried figs
- 1 cup orange juice
- ½ cup maple syrup
- ¼ cup molasses
- 1½ teaspoons cinnamon

1. Peel carrots and potatoes. Slice ¼ inch thick. Boil or steam until tender; drain.

2. Cut figs in halves or quarters, depending upon size of fig used. In a large bowl, combine all ingredients.

3. Pour into a large greased casserole or baking dish. Bake, covered, in a 350° oven for 30 minutes. Uncover and bake 15 minutes longer.

Yield: 8 servings.

Calcium: 117 mg. per serving.

Calories: 221 per serving.

CORN PUDDING WITH CHEESE

- 2 cups White Sauce (recipe follows)
- 1 cup grated Cheddar or Monterey Jack cheese
- 4-ounce can mild chilies, diced
- 1 teaspoon maple syrup or honey
- ½ teaspoon cumin
- 2 eggs, lightly beaten
- 2 16-ounce cans corn, drained
- Paprika

1. Prepare the White Sauce. Add cheese and stir until melted. Add chilies, maple syrup, and cumin. Stir in eggs. Blend in corn.

2. Pour into lightly greased casserole. Sprinkle with paprika. Bake, uncovered, in a 350° oven for 35 minutes.

Yield: 8 servings.

Calcium: 200 mg. per serving.

Calories: 282 per serving.

WHITE SAUCE

- 2 tablespoons margarine
- 2 tablespoons all-purpose flour
- 2 cups skim milk
- Salt and pepper to taste

1. Melt margarine in saucepan and stir in flour. Cook for 2 minutes.

2. Add milk slowly and cook until thickened, stirring constantly. Season to taste.

Yield: 2 cups.

Calcium: 175 mg. per ½ cup.

Calories: 102 per ½ cup.

SALADS

Many of these salads double as main dishes; others are intended to accompany pasta, cheese, or seafood dishes. Included are three recipes using kelp, which is loaded with calcium, but not all of it is absorbed because of the high fiber content of all seaweed. Kelp is also *very* high in iodine, potassium, and vitamin K. Although these are essential nutrients, excessive amounts can cause problems. Therefore, kelp should not be eaten in large amounts. The recipes are included here for variety—they are not recommended as mainstays in the diet. Kelp can be found in most Oriental markets, and sometimes in health-food stores. It is often called wakame.

LUNCHEON SARDINE SALAD

- 3½-ounce can sardines, drained
- 2 teaspoons fresh lemon juice
- ½ teaspoon garlic powder
- 1 scallion, chopped
- 1 medium tomato, diced
- 1 cup torn spinach leaves, washed

Mash sardines with lemon juice and garlic powder. Stir in scallions and tomato. Serve on a bed of spinach.

Yield: 1 serving.

Calcium: 375 mg. per serving.

Calories: 260 per serving.

Variation: Hollow out tomato and stuff with sardine mixture.

SALMON NIÇOISE

- 15½-ounce can kidney beans, rinsed and drained
- 6 ounces olives, drained
- 2 tomatoes, chopped
- 1 small red onion, cut in half and sliced
- ¼ cup chopped fresh parsley
- 1 small head red leaf or romaine lettuce
- 2 7-ounce cans salmon, drained
- 2 ounces almonds, sliced

DRESSING:

- 3 tablespoons cider vinegar or lemon juice
- 2 tablespoons oil
- 2 garlic cloves, minced
- 1 teaspoon chopped capers (optional)*
- ½ teaspoon Dijon mustard

1. Place beans, olives, tomato, red onion, and parsley in a bowl. Whisk together the dressing ingredients and pour half over the bean mixture.

2. Wash and dry lettuce; tear into bite-size pieces. Arrange on a serving platter. Mound vegetable mixture in the center.

3. Break salmon into chunks. Place salmon around outside of vegetables. Sprinkle almonds over all and pour remaining dressing over all.

Yield: 6 servings.

Calcium: 236 mg. per serving.

Calories: 284 per serving.

* Not included in calorie and calcium calculations.

SALMON AND SPINACH SALAD

2 7-ounce cans salmon, drained and flaked
1 pound cooked potatoes, peeled and cubed
10 ounces fresh spinach leaves, washed and torn
1 cup sliced fresh mushrooms
1 cup thinly sliced carrots
1 cup raw sliced broccoli or cauliflower
Spicy Dill and Yogurt Dressing (see Index)

1. Do not remove bones from salmon. They are easy to chew and contain much calcium.

2. Combine all ingredients and toss with Spicy Dill and Yogurt Dressing.

Yield: 4 servings.

Calcium: 308 mg. per serving.

Calories: 240 per serving.

SALMON AND VEGETABLE PASTA SALAD

2 cups elbow macaroni, cooked, drained and rinsed
2 cups chopped broccoli, steamed
14 ounces canned salmon, drained and broken into chunks
1½ cups diced celery
½ cup cubed Cheddar cheese
½ cup chopped green pepper
½ cup mayonnaise
¼ cup chopped sweet pickles
2 tablespoons honey
Salad greens, washed and drained

1. In a large bowl combine macaroni, broccoli, salmon, celery, cheese, and green pepper.

2. In a small bowl, blend the mayonnaise, pickles, and honey. Toss with macaroni mixture.

3. Line bowls with salad greens. Spoon salad into bowls.

Yield: 6 servings.

Calcium: 271 mg. per serving.

Calories: 336 per serving.

SMOKED FISH SALAD

7 ounces kippers or smoked baby clams
2 cups shredded green cabbage
1 medium new potato, boiled and finely sliced
1 cup chopped red pepper
2 scallions, chopped
1 teaspoon caraway seeds
Creamy Salad Dressing (see Index)

Cut kippers or clams into bite-size pieces and combine with remaining ingredients.

Yield: 4 side-dish servings; 2 entrée servings.

Calcium: 71 mg. per side-dish serving; 142 per entrée serving.

Calories: 164 per side-dish serving; 328 per entrée serving.

KIPPER SALAD WITH HOT PEPPER DRESSING

- 3½-ounce can kippered herring steaks, drained
- 2 scallions, chopped
- 2 tomatoes, chopped
- 1 cup chopped cucumbers
- ¼ cup slivered almonds
- 1 tablespoon mayonnaise
- 2 cups coarsely chopped fresh spinach leaves
- Juice of 1 lime
- Hot pepper sauce (e.g., Tabasco) to taste

1. In a large bowl, combine kippers, scallions, tomatoes, cucumbers, almonds, and mayonnaise.

2. Add spinach and toss.

3. Sprinkle with lime juice. Season with hot pepper sauce.

4. Toss well and serve.

Yield: 4 servings.

Calcium: 97 mg. per serving.

Calories: 172 per serving.

SCALLOPS AND KELP

8 sea scallops or 16 bay scallops
5 ounces kelp (wakame)

MISO MAYONNAISE:

½ cup mayonnaise
2 tablespoons miso (fermented bean curd)
Hot Chinese mustard to taste*

GARNISHES:

¼ lemon, thinly sliced (optional)*
3 tablespoons finely chopped parsley

1. Poach scallops and, if using sea scallops, slice in half horizontally.

2. Place kelp in strainer, pour hot water over, then immediately run it under cold water. Chop into small pieces and blend with scallops.

3. Combine mayonnaise ingredients and spoon over scallops. Garnish with lemon and parsley.

Yield: 2 servings.

Calcium: 424 mg. per serving.

Calories: 135 per serving (with one tablespoon of dressing at 86 calories per tablespoon).

* Not included in calorie and calcium calculations.

BEAN SPROUT, KELP, AND TOMATO SALAD

5 ounces kelp (wakame)
5 ounces bean sprouts
2 medium tomatoes, sliced

Soy Sauce Dressing:

¼ cup vinegar
¾ cup salad oil
3 tablespoons soy sauce
Pepper to taste

1. Place kelp in strainer, pour hot water over, then immediately run it under cold water. Drain and cut into bite-size pieces.

2. Blanch bean sprouts and drain.

3. Place tomato slices on serving dish. Layer bean sprouts, then kelp on top. Combine all dressing ingredients and mix well. Pour over salad just before serving.

Yield: 4 servings.

Calcium: 410 mg. per serving.

Calories: 115 per serving (includes 1 tablespoon of dressing at 81 calories per tablespoon).

ORIENTAL SALAD

1 head cauliflower (approximately 1½ pounds)
1 bunch broccoli (approximately 1½ pounds)
¼ cup vegetable oil
¼ cup vinegar
2 tablespoons soy sauce
2 tablespoons sugar
2 large garlic cloves, minced
1 teaspoon powdered ginger

1. Trim and wash vegetables. Break into florets (save broccoli stems for another use), and steam until tender but still crisp.

2. Combine remaining ingredients in a separate bowl, and toss with vegetables.

Yield: 8 servings.

Calcium: 116 mg. per serving.

Calories: 199 per serving.

ORIENTAL KELP AND CUCUMBER SALAD

1 small cucumber, or 2 Kirby cucumbers, peeled and thinly sliced
1 cup water, with 1 teaspoon salt
2 ounces kelp (wakame)
3 tablespoons rice vinegar
1 tablespoon sugar
Pinch of salt
1 teaspoon soy sauce
3 tablespoons water
Garnish: fresh ginger, shredded or sliced razor thin (optional)*

1. Soak sliced cucumbers in salted water for 5–10 minutes. Drain and squeeze out extra water.

2. Wash kelp and remove tough parts. Cut into 1-inch pieces.

3. Place kelp in strainer and pour hot water over. Then immediately run under cold water. Drain.

4. Combine rice vinegar, sugar, salt, soy sauce, and 3 tablespoons water to make dressing.

5. Place cucumbers and kelp in serving bowl. Pour dressing over. Garnish with shredded ginger, if desired.

Yield: 4 servings.

Calcium: 164 mg. per serving.

Calories: 20 per serving.

* Not included in calories and calcium calculations.

ORANGE AND ONION SALAD

4 large navel oranges, peeled
1 medium red onion, peeled and sliced thin
Spicy Dill and Yogurt Dressing (see Index)
1 medium head Boston lettuce

1. Remove white membrane and cut oranges into thin slices.

2. In a bowl, toss orange slices with onion slices. Add dressing and toss again.

3. Refrigerate until chilled.

4. Serve on Boston lettuce.

Yield: 4 servings.

Calcium: 99 mg. per serving (with 1 tablespoon dressing).

Calories: 118 per serving (with 1 tablespoon dressing).

ORANGE AND BROCCOLI PASTA SALAD

- 1 pound pasta shells
- 1/4 cup shelled sunflower seeds
- 2 tablespoons sesame oil
- 1/4 cup vinegar
- 1 tablespoon soy sauce
- 1/2 teaspoon powdered ginger
- 2 cups chopped fresh broccoli
- 2 oranges, peeled and cubed

1. Cook pasta until tender, but still chewy. Drain.

2. Sauté seeds in oil until slightly browned. Add vinegar, soy sauce, and ginger.

3. Stir in broccoli and oranges.

4. Toss with pasta and serve at room temperature.

Yield: 8 servings.

Calcium: 35 mg. per serving.

Calories: 207 per serving.

POACHED ORANGE SALAD

6 navel oranges, peeled
1 tablespoon grated orange rind
⅔ cup orange juice
⅔ cup water
⅔ cup honey
1 cup low-fat plain yogurt
Lettuce leaves

1. Remove white membrane from oranges and slice.

2. Put orange rind in saucepan with orange juice, water, and honey. Simmer for 10 minutes, stirring frequently.

3. Add oranges and simmer for 2 minutes more. Remove oranges from liquid and chill. Strain liquid.

4. Blend 2 tablespoons of liquid with the yogurt. Line 6 salad plates with lettuce leaves. Arrange chilled orange slices on top of lettuce. Spoon yogurt mixture over oranges.

Yield: 6 servings.

Calcium: 149 mg. per serving.

Calories: 233 per serving.

ORANGE AND CRANBERRY DELIGHT

1 cup chopped raw cranberries
2 tablespoons sugar
1 apple, cored and diced
2 navel oranges, peeled and sliced
½ cup diced celery
2½ cups Orange Cheese Dressing (see Index)
Lettuce leaves

1. Place cranberries in a bowl and sprinkle with sugar.

2. Combine apple, oranges, and celery in a larger bowl. Add cranberries and dressing; mix together and chill.

3. Line salad bowl with lettuce and place fruit salad in bowl.

Yield: 6 servings.

Calcium: 106 mg. per serving.

Calories: 161 per serving.

VEGETABLE SALAD WITH LEMON DRESSING

- 1 medium cucumber, peeled and sliced
- 10 ounces spinach, washed and patted dry
- 1 cup chopped raw cauliflower
- 2 medium tomatoes, cubed
- ½ cup chopped almonds
- ½ cup grated carrots

Lemon Dressing:

- 5 tablespoons fresh lemon juice
- 3 tablespoons oil
- ½ teaspoon Dijon mustard
- ¼ teaspoon dried oregano
- ¼ teaspoon dried thyme

1. Place cucumbers in a bowl and toss with Lemon Dressing, which is prepared by mixing lemon juice, oil, mustard, oregano, and thyme. Cover and let stand for 2 hours.

2. Remove stems from spinach and tear into bite-size pieces. Combine with cauliflower, tomatoes, almonds, and carrots.

3. Just before serving, pour cucumber slices with dressing over salad. Toss.

Yield: 6 servings.

Calcium: 100 mg. per serving.

Calories: 173 per serving.

SPINACH SALAD WITH ORANGE

10 ounces spinach
2 oranges, peeled and sectioned
3 hard-boiled eggs, sliced
2 hard-boiled egg whites, sliced
½ cup sunflower seeds
½ medium red onion
Sesame Dressing (see Index)

1. Clean spinach. Dry and tear into bite-size pieces.
2. Add oranges, eggs, and seeds.
3. Peel, slice and separate red onion, add to spinach.
4. Serve with Sesame Dressing.

Yield: 4 servings.

Calcium: 151 mg. per serving.

Calories: 292 per serving.

MEDITERRANEAN SPINACH SALAD

10 ounces spinach, torn and washed
1 cup mandarin orange segments
1 cup shredded red cabbage
½ cup cooked garbanzo beans
10 black olives
4 ounces feta cheese, crumbled
1 small cucumber, thinly sliced

DRESSING:

2 tablespoons oil
4 tablespoons rice vinegar or lime juice
½ teaspoon Italian seasoning
½ teaspoon garlic powder

1. Combine salad ingredients in a large bowl.
2. In a separate bowl, whisk dressing ingredients.
3. Just before serving, stir dressing and toss with spinach salad.

Yield: 4 servings.

Calcium: 276 mg. per serving.

Calories: 288 per serving.

SPINACH SALAD WITH CHEESE

10 ounces spinach, washed and drained
1 cup cooked garbanzo beans
1 cup raw broccoli cut into bite-size pieces
4-ounce can artichoke hearts, in water
2 ounces part-skim mozzarella cheese, shredded
2 hard-boiled eggs
½ avocado, cubed (optional)
Creamy Salad Dressing (see Index)

1. Tear spinach leaves into bite-size pieces.
2. Combine remaining ingredients and mix with spinach.
3. Dress with Creamy Salad Dressing.

Yield: 4 servings.

Calcium: 255 mg. per serving.

Calories: 252 per serving.

VEGETABLE TERIYAKI SALAD

- 1 small head cauliflower, cut into florets
- 1 small bunch broccoli, cut into florets
- 1 tablespoon sesame oil
- 3 garlic cloves, minced
- 1 pound firm tofu, thinly sliced
- 1 cup sliced fresh mushrooms
- ½ cup sliced almonds
- ½ cup orange juice
- ¼ cup oil
- ¼ cup dry sherry
- ¼ cup soy sauce
- 2 tablespoons sugar
- 1 tablespoon vinegar
- ½ teaspoon ginger

1. Steam cauliflower and broccoli until tender but still crisp.

2. Heat sesame oil and sauté garlic, then add tofu slices and sauté 5 minutes.

3. Combine cauliflower, broccoli, tofu mixture, mushrooms, and almonds.

4. Prepare a dressing with the remaining ingredients. Toss with vegetables.

Yield: 6 servings.

Calcium: 290 mg. per serving.

Calories: 347 per serving.

MAPLE-NUT CARROT SALAD

- 1 cup chopped dried black figs
- 1 tablespoon oil
- ½ cup slivered or sliced almonds
- ½ cup shelled sunflower seeds
- 2 tablespoons sesame seeds
- ½ cup mayonnaise
- 2 tablespoons maple syrup
- 1½ pounds carrots, shredded

1. Soak figs in hot water for 5 minutes; drain.

2. Heat oil in a skillet. Add almonds, sunflower and sesame seeds; cook until lightly browned.

3. Stir in mayonnaise and maple syrup. Toss with carrots. Serve chilled.

Yield: 10 servings.

Calcium: 89 mg. per serving.

Calories: 286 per serving.

PASTA PRIMAVERA SALAD WITH CHEESE

8 ounces macaroni shells or twists, cooked and drained
16 ounces ricotta or low-fat cottage cheese
1/2 cup shredded carrots
1/4 cup finely chopped scallions
2 tablespoons bottled low-calorie Italian dressing
1/8 teaspoon fennel seed
1/8 teaspoon pepper
1/2 cup chopped broccoli
1/2 cup chopped red pepper

1. Combine cooked macaroni with cheese, carrots and scallions.

2. Combine dressing with fennel and pepper.

3. Toss salad with dressing. Sprinkle with broccoli and red pepper.

Yield: 6 servings.

Calcium: 75 mg. per serving.

Calories: 254 per serving.

PASTA AND BEANS SALAD

- 1 cup pasta shells, cooked
- 15-ounce can tomatoes, drained and chopped
- 1 cup cooked kidney beans, drained
- 1 cup cooked navy beans, drained
- 1 cup cooked broccoli florets
- 1 teaspoon dried basil
- ½ teaspoon garlic powder

1. Combine all ingredients and cook until heated.
2. Serve hot as a side dish or chilled as a pasta salad.

Yield: 6 servings.

Calcium: 94 mg. per serving.

Calories: 207 per serving.

MIDDLE EASTERN BEAN SALAD

- 3 cups canned kidney beans, drained
- 1½ cups canned garbanzo beans, drained
- 1 cup finely chopped parsley
- ½ head of cauliflower, chopped
- 1 medium red onion, halved and sliced
- 1 large tomato, diced
- 1 cucumber, diced
- ½ cup oil
- ⅓ cup lemon juice
- 2 tablespoons vinegar
- 1½ teaspoons Italian seasoning
- 1 teaspoon salt
- ½ teaspoon ground pepper

1. Combine all the beans and vegetables in a large bowl.
2. In a small bowl, mix dressing ingredients.
3. Toss salad with dressing.

Yield: 8 servings.

Calcium: 174 mg. per serving.

Calories: 334 per serving.

TOFU AND CABBAGE SALAD

½ tablespoon sesame seeds
2 tablespoons sesame or peanut oil
8 ounces firm tofu, cubed
⅛ teaspoon crushed red pepper
2 cups shredded Chinese cabbage
1 cup shredded red cabbage
1 cup cooked garbanzo beans
1 cup bean sprouts
4-ounce jar artichoke hearts, in water
2 celery ribs, sliced
6 tablespoons rice vinegar

1. Sauté sesame seeds in oil until seeds are browned. Add tofu and crushed red pepper; sauté 5 minutes longer.

2. Combine remaining ingredients except vinegar and toss well with tofu.

3. Add vinegar and toss again.

Yield: 6 servings.

Calcium: 64 mg. per serving.

Calories: 171 per serving.

CHINESE COLESLAW

- 6 cups shredded cabbage
- ½ cup chopped scallions
- 11-ounce can mandarin oranges, drained, or 2 small navel oranges, peeled and sectioned
- ½ cup sliced almonds
- ½ cup vegetable oil
- 2 tablespoons vinegar
- 2 tablespoons sugar
- 1 tablespoon sesame seeds
- 1 tablespoon soy sauce
- 2 teaspoons mayonnaise
- ½ teaspoon salt
- ¼ teaspoon paprika
- ⅛ teaspoon dry mustard

1. Combine cabbage, scallions, oranges, and almonds.

2. In a separate bowl, mix the remaining ingredients to form the dressing.

3. Just before serving, stir dressing and toss with cabbage salad.

Yield: 8 servings.

Calcium: 69 mg. per serving.

Calories: 245 per serving.

ALMOND COLESLAW

- 2 cups shredded red cabbage
- 2 cups shredded green cabbage
- 5 scallions, chopped
- 2 celery ribs, thinly sliced
- 1 red pepper, chopped
- ½ cucumber, diced
- 1 cup sliced almonds

DRESSING:

- ⅓ cup mayonnaise
- 2 tablespoons rice or cider vinegar
- ½ teaspoon salt
- ⅛ teaspoon ground pepper

1. In a large bowl, combine cabbages, scallions, celery, red pepper, and cucumber.

2. In a separate bowl, prepare dressing by mixing mayonnaise, vinegar, salt, and pepper.

3. Toss dressing with salad; top with sliced almonds.

Yield: 8 servings.

Calcium: 83 mg. per serving.

Calories: 199 per serving.

PEANUT BUTTER COLESLAW

- 3/4 cup oil
- 1/3 cup vinegar
- 1 tablespoon peanut butter
- 1 tablespoon soy sauce
- 5 garlic cloves, minced
- 1 1/2 teaspoons cumin
- 1 1/2 teaspoons sugar
- Cayenne pepper
- 1 small head cabbage, shredded
- 1 carrot, shredded
- 1/2 large onion, minced
- 1/2 cup sliced almonds, toasted

1. Combine oil, vinegar, peanut butter, soy sauce, and seasonings to prepare dressing.

2. Combine cabbage, carrot, onion, and almonds. Toss with dressing.

Yield: 8 servings.

Calcium: 65 mg. per serving.

Calories: 282 per serving.

Vegetables

Many vegetables are good sources of calcium, and almost any vegetable dish can be transformed into a high-calcium one by adding cheese or other dairy products. Even so, you should not rely solely on vegetables for calcium. Although many have large amounts of calcium, it is not as readily absorbed as that found in dairy products and other foods. Having noted this, we can add that the following recipes provide combinations designed to increase calcium uptake. Plus they are loaded with other important nutrients and are delicious to eat.

VEGETABLES PROVENÇALE

- 2 medium zucchini, cut into chunks (approximately 2 cups)
- 1 onion, halved and sliced
- 1½ cups cauliflower florets
- 1 cup tomato sauce
- ½ cup chopped stewed tomatoes
- 1 teaspoon dried basil
- 1½ cups chopped cabbage

1. Place zucchini, onion, cauliflower, tomato sauce, tomatoes, and basil in a saucepan; simmer until vegetables are tender but still crisp.
2. Stir in cabbage and cook 5 minutes longer.

Yield: 6 servings.

Calcium: 40 mg. per serving.

Calories: 43 per serving.

STIR-FRIED VEGETABLES WITH TOFU

- 1 tablespoon vegetable oil
- 4 garlic cloves, minced
- ½ pound snow peas, trimmed
- 1 sweet red pepper, cubed
- ¼ pound mushrooms, sliced
- ¾ cup sliced scallions
- 14 ounces firm tofu, drained and cubed
- 2 tablespoons water
- 2 teaspoons cornstarch
- 1 tablespoon sesame oil
- 1½ teaspoons soy sauce
- ½ teaspoon vinegar
- ½ teaspoon sugar
- ¼ teaspoon ground pepper

1. Heat oil in large skillet or wok. Stir-fry garlic, then peas and red pepper, then mushrooms, until tender-crisp.

2. Add scallions and tofu; cook 1 minute. Remove from heat.

3. Prepare sauce. Combine water and cornstarch to form a paste; add remaining ingredients.

4. Pour sauce into skillet and cook until thickened and heated.

Yield: 6 servings.

Calcium: 108 mg. per serving.

Calories: 144 per serving.

PICKLED VEGETABLES IN MUSTARD-GINGER MARINADE

- 1 cup dry white wine
- 1 cup vinegar
- ½ cup water
- ¼ cup brown sugar
- 2 teaspoons salt
- 1 teaspoon mustard seeds
- 1 teaspoon powdered ginger
- 1 teaspoon ground allspice
- 1 medium cauliflower, washed and chopped
- 2 medium cucumbers, peeled and cubed
- 2 medium onions, sliced
- 2 tablespoons flour

1. Combine wine, vinegar, water, sugar, salt, and spices in a large pot; boil for 5 minutes.

2. Add vegetables; lower heat and simmer until tender but still crisp.

3. Mix flour with a little water to form a paste. Stir in some hot marinade, then pour back into pot, stirring until slightly thickened.

4. Remove from heat; chill.

Yield: 6 servings.

Calcium: 66 mg. per serving.

Calories: 117 per serving.

MARINATED VEGETABLES

- 1 cup wine vinegar
- ½ cup vegetable oil
- ⅓ cup sugar
- 2 teaspoons dried basil
- 1 teaspoon salt
- ½ teaspoon ground pepper
- 2 cups sliced cucumber
- 1 cup thinly sliced carrots
- 1 cup chopped fresh broccoli florets
- 1 cup chopped fresh cauliflower florets
- 6 medium tomatoes, cut into wedges
- 2 medium onions, thinly sliced
- 1 cup sliced almonds, toasted

1. Combine marinade by mixing vinegar, oil, sugar. basil, salt, and pepper.

2. Place all the vegetables in a large bowl. Pour marinade over and chill for 3–4 hours, stirring from time to time.

3. Just before serving, drain and sprinkle with almonds.

Yield: 8 servings.

Calcium: 103 mg. per serving.

Calories: 252 per serving, if drained.

SUNFLOWERY VEGETABLES

- 1 tablespoon oil
- ½ cup shelled sunflower seeds
- 1 cup diced carrots
- 1 cup diced celery
- ½ cup water
- 2 cups chopped greens (kale, turnips, etc.)
- ½ teaspoon dried oregano
- Juice of ½ lemon

1. Sauté seeds in oil until lightly colored. Add carrots and celery, stirring until well coated.

2. Add water and simmer 5 minutes.

3. Add greens and oregano. Cook 2 minutes longer.
4. Remove from heat. Squeeze lemon juice over, stir and serve.

Yield: 4 servings.

Calcium: 212 mg. per serving.

Calories: 240 per serving.

VEGETABLE CURRY

1 tablespoon oil
1 small onion, chopped
2 garlic cloves, minced
3 medium tomatoes, chopped
2 teaspoons curry powder
½ teaspoon salt
⅛ teaspoon ground pepper
1 cup coarsely chopped broccoli
1 cup coarsely chopped cauliflower
1 cup halved brussels sprouts

1. Heat oil and sauté onion and garlic until limp. Add tomatoes, curry powder, salt, and pepper. Simmer 15 minutes until thickened to a sauce.
2. Meanwhile steam broccoli, cauliflower and brussels sprouts until tender but still crisp.
3. Stir cooked vegetables into curry sauce and cook 5 minutes longer, adding a small amount of water if necessary. Serve as a side dish with rice or other grain.

Yield: 4 servings.

Calcium: 75 mg. per serving.

Calories: 103 per serving.

MASHED PARSNIPS

2 pounds fresh parsnips, peeled, quartered and steamed
1/2 cup skim milk
2 tablespoons lemon juice
1 teaspoon salt
1/2 teaspoon nutmeg
Fresh parsley, minced

1. Mash parsnips with milk.
2. Stir in lemon juice, salt, and nutmeg. Reheat if necessary.
3. Garnish with parsley.

Yield: 6 servings.

Calcium: 98 mg. per serving.

Calories: 110 per serving.

OKRA AND TOMATOES

1 small onion, chopped
2 garlic cloves, minced
1 tablespoon oil
1/2 pound plum or Italian canned tomatoes, chopped
1/2 teaspoon salt
1/2 teaspoon dried basil
1/8 teaspoon ground pepper
2 cups chopped greens (kale, turnips, etc.)
1 pound okra, chopped

1. Sauté onion and garlic in oil until lightly browned. Stir in tomatoes, salt, basil, and pepper; cook approximately 10 minutes, or until tomatoes are softened.
2. Add greens and okra. Cover and simmer over low heat until okra is tender. You may need to add a small amount of water.

Yield: 6 servings.

Calcium: 166 mg. per serving.

Calories: 75 per serving.

CUCUMBER TOMATO RAITA

1 cucumber, peeled and grated
½ small onion, grated
1 cup low-fat plain yogurt
1 tomato, diced
1 tablespoon chopped fresh mint
¼ teaspoon salt

Combine all the ingredients and chill. Serve as an accompaniment to spicy foods.

Yield: 4 servings.

Calcium: 120 mg. per serving.

Calories: 105 per serving.

STIR-FRIED CABBAGE

1 tablespoon oil
½ onion, sliced
¼ cup sliced almonds
1½ tablespoons sesame seeds
1 green apple, peeled and thinly sliced
2 cups coarsely chopped cabbage (any type)
Juice of ½ orange
Salt to taste

1. Heat oil in a skillet. Quickly stir-fry the onions, almonds, and seeds.

2. Add apple and sauté till softened slightly.

3. Stir in cabbage; cook until cabbage is tender. Sprinkle with juice. Season with salt.

Yield: 2 generous servings.

Calcium: 120 mg. per serving.

Calories: 310 per serving.

EGGPLANT IN LIGHT CHEESE SAUCE

- 2 tablespoons oil
- 1 small onion, chopped
- 2 cups eggplant cubes
- 8 ounces low-fat plain yogurt
- 1 tablespoon dark molasses
- 1/4 cup dry white wine
- 1 teaspoon Italian seasoning
- 2 ounces mozzarella or muenster cheese, shredded

1. Sauté onion in oil until soft. Add eggplant and cook until soft, adding a little water if necessary, to prevent sticking.

2. Stir together yogurt and molasses until light. Add to the saucepan along with wine and seasoning.

3. Simmer for 10 minutes. Just before serving, sprinkle with cheese.

Yield: 2 servings.

Calcium: 467 mg. per serving.

Calories: 508 per serving.

MUSHROOM AND TOFU PAPRIKASH

- 1 medium onion, diced
- 3 garlic cloves, finely minced
- 2 tablespoons margarine
- 3/4 pound fresh mushrooms
- 1/2 pound firm tofu, diced
- 1 tablespoon snipped fresh dill, or 1 teaspoon dried dillweed
- 1 teaspoon salt
- 1/2 teaspoon paprika
- 1/8 teaspoon ground pepper
- 1 cup low-fat plain yogurt

1. Sauté onion and garlic in margarine until softened.

2. Wipe mushrooms with a damp cloth. Cut off stems and reserve for some other use (good in a stuffing for baked vegetables). Slice mushrooms, then sauté in margarine until softened.

3. Add tofu and seasonings; sauté for 5 minutes, stirring occasionally.

4. Stir in yogurt and cook until thoroughly heated (do not boil).

Yield: 4 servings.

Calcium: 217 mg. per serving.

Calories: 204 per serving.

HOT AND PEPPERY ZUCCHINI

- ½ cup chopped onion
- 2 garlic cloves, chopped
- 1 tablespoon oil
- ¾ pound zucchini, cubed
- ⅓ cup diced celery
- 3 ounces part-skim mozzarella cheese, diced
- ½ cup canned stewed tomatoes, mashed
- ½ teaspoon crushed red pepper

1. Sauté onion and garlic in oil until onion is transparent; add zucchini and celery. Cook until slightly softened.

2. Transfer to a baking dish. Add cheese and bake 10 minutes at 350°.

3. Combine tomatoes and crushed red pepper in a saucepan and stew 10 minutes.

4. Pour over zucchini mixture.

Yield: 4 servings.

Calcium: 195 mg. per serving.

Calories: 139 per serving.

TOMATOES WITH ALMOND-CHEESE STUFFING

4 large tomatoes
1 tablespoon margarine
¼ cup chopped almonds
⅓ cup seasoned bread crumbs
¼ cup grated Parmesan cheese

1. Cut tomatoes in half; scoop out pulp.

2. Drain liquid and seeds from pulp. Chop pulp and set aside.

3. Melt margarine in a saucepan. Sauté almonds until lightly browned. Stir in bread crumbs, cheese, and chopped pulp.

4. Stuff mixture into tomato shells.

5. Bake 20–25 minutes in a 375° oven.

Yield: 4 servings.

Calcium: 130 mg. per serving.

Calories: 174 per serving.

CHINESE BRUSSELS SPROUTS

1 tablespoon sesame oil
1 onion, chopped
3 garlic cloves, minced
2 tablespoons sesame seeds
Small can (8 ounces) bamboo shoots
1 pint brussels sprouts, steamed

1. Heat oil and sauté onion, garlic, and seeds until lightly browned.

2. Stir in bamboo shoots and brussels sprouts; heat thoroughly.

Yield: 4 servings.

Calcium: 59 mg. per serving.

Calories: 141 per serving.

BRUSSELS SPROUTS AND CAULIFLOWER WITH NUT SAUCE

3 tablespoons margarine
2 garlic cloves, minced
½ cup ground almonds or hazelnuts
3 tablespoons flour
2 cups low-fat plain yogurt
3 tablespoons grated Parmesan cheese
Dash of pepper
1 pint brussels sprouts, steamed
2 cups cauliflower florets, steamed

1. Melt magarine; sauté garlic, then nuts until lightly browned. Add flour and cook 2 minutes, stirring constantly.

2. Stir in yogurt, Parmesan, and pepper. Cook about 2 minutes.

3. Pour over steamed vegetables.

Yield: 6 servings.

Calcium: 242 mg. per serving.

Calories: 251 per serving.

HOT AND SPICY CAULIFLOWER

1 head cauliflower
½ cup tomato sauce
1 medium tomato, chopped
2 tablespoons chopped green olives, or 1 tablespoon chopped capers
1 teaspoon crushed red pepper
½ teaspoon dried oregano

1. Separate cauliflower into florets and steam until tender. Set aside.

2. Combine remaining ingredients in a saucepan and simmer 30 minutes.

3. Pour over cauliflower, heat and serve.

Yield: 4 servings.

Calcium: 76 mg. per serving.

Calories: 97 per serving.

BROCCOLI AND CAULIFLOWER IN GARLIC NUT SAUCE

1 medium head cauliflower
1 medium bunch broccoli
¼ cup vegetable oil
4 garlic cloves, minced
2 tablespoons sesame seeds
½ cup blanched and slivered almonds
¼ cup fresh lemon juice
Salt and pepper to taste

1. Remove core and tough stalks from vegetables and cut into florets. Steam vegetables in water until tender yet still crunchy. Set aside.

2. Meanwhile heat oil in a skillet. Add garlic, seeds, and almonds; cook until lightly browned. Stir in lemon juice, salt, and pepper.

3. Pour over vegetables. Refrigerate for 1 hour or more. Bring to room temperature before serving.

Yield: 8 servings.

Calcium: 114 mg. per serving.

Calories: 186 per serving.

BROCCOLI VINAIGRETTE

3 pounds broccoli
½ cup oil
2 tablespoons vinegar
1 teaspoon Dijon mustard
1 garlic clove, minced
Salt and pepper to taste
2 teaspoons dried dillweed
1 tablespoon finely chopped parsley
1 hard-boiled egg, finely chopped

1. Wash and trim broccoli; separate into stalks. Steam until tender but still crisp; drain.

2. Combine oil, vinegar, mustard, garlic, salt, pepper, and dillweed. Pour over broccoli and chill several hours.

3. Just before serving, garnish with parsley and egg.

Yield: 8 servings.

Calcium: 190 mg. per serving.

Calories: 194 per serving.

Note: This dressing works well with other steamed vegetables such as cauliflower, string beans, carrots, and zucchini.

CREAMED SPINACH

10-ounce package frozen chopped spinach
1 tablespoon margarine
1 tablespoon flour
¾ cup skim milk
½ teaspoon garlic powder
1 tablespoon mayonnaise

1. Steam spinach.

2. Melt margarine, stir in flour, and cook 2 minutes. Stir in milk and garlic; cook until thickened. Stir in mayonnaise. Pour over spinach, mix and serve.

Yield: 2 servings.

Calcium: 280 mg. per serving.

Calories: 189 per serving.

Sauces, Marinades, and Dressings

For many people, it's the dressing or sauce that makes the dish. And for many calorie-conscious people, sauces and dressings are their downfall. Most of us think of a salad or vegetable dish as being so low in calories that we hardly have to count them. But a 100-calorie salad can be turned into a 1,000-calorie extravaganza by dousing it with a rich dressing and then topping it with bacon bits or other high-calorie salad-bar extras. The sauces, dressings, and marinades in this section are designed to add flavor and varying amounts of calcium with a minimum of calories.

SPICY DILL AND YOGURT DRESSING

1 cup low-fat plain yogurt
2 tablespoons mayonnaise
1/4 cup finely chopped scallions
2 teaspoons Dijon mustard
1 teaspoon dried dillweed

Combine all ingredients.

Yield: 1 1/4 cups.

Calcium: 23 mg. per tablespoon.

Calories: 18 per tablespoon.

Note: This dressing goes well with poached fish or fish salads.

ORANGE CHEESE DRESSING

1 cup low-fat cottage cheese
1 cup low-fat plain yogurt
1/2 cup orange juice
4 tablespoons honey
1/4 teaspoon powdered ginger
1/4 teaspoon nutmeg
1/2 teaspoon cinnamon

1. Place cottage cheese and yogurt in a mixing bowl or blender. Beat or whirl at medium speed.

2. Gradually add orange juice, honey, and spices. Beat or whirl until smooth. Chill before serving.

Yield: 2 1/2 cups.

Calcium: 254 mg. per cup.

Calories: 274 per cup.

Note: This is particularly good with Orange and Cranberry Delight (see Index).

AVOCADO SALAD DRESSING

1 avocado, peeled and cubed
1 cup skim milk
3 tablespoons grated Parmesan cheese
2 rounded tablespoons chopped scallions
2 garlic cloves, minced
4 drops hot pepper sauce (e.g., Tabasco)
½ teaspoon chili powder
¼ teaspoon salt
Generous sprinkling of pepper

1. In a medium bowl, use a hand mixer to mash avocado with milk until smooth.

2. Add remaining ingredients and mix until well blended.

Yield: 1½ cups.

Calcium: 11 mg. per tablespoon.

Calories: 22 per tablespoon.

Note: This can also be a dip by reducing the amount of milk to ¾ cup.

TOFU DILL DRESSING

8 ounces soft tofu
¼ cup fresh lemon juice
2 tablespoons oil
1 tablespoon grated onion
2 teaspoons snipped fresh dill, or 1 teaspoon dried dillweed
½ teaspoon salt

Place all ingredients in a blender or food processor and whirl until smooth. Serve with green salads.

Yield: 1 cup.

Calcium: 20 mg. per tablespoon.

Calories: 28 per tablespoon.

LIME AND SESAME SEED DRESSING

½ cup oil
¼ cup fresh lime juice
2 tablespoons sesame seeds, well toasted
2 tablespoons molasses or maple syrup (optional)*

Combine all ingredients. If a sweeter dressing is desired, include molasses or maple syrup. Use as a dressing on fruit salads or other salads containing fruit.

Yield: ¾ cup.

Calcium: 3 mg. per tablespoon.

Calories: 99 per tablespoon.

* Not included in calorie and calcium calculations. Maple syrup and light molasses have 50 calories and 33 mg. calcium per tablespoon.

CREAMY SALAD DRESSING

1¾ cups low-fat cottage cheese
1 cup skim milk
2 scallions, chopped
1 teaspoon garlic powder
1 teaspoon dried dillweed
¼ teaspoon pepper

Whirl all the ingredients in a blender until smooth and creamy.

Yield: 2½ cups.

Calcium: 14 mg. per tablespoon.

Calories: 10 per tablespoon.

SESAME DRESSING

- 1/4 cup sesame seeds
- 1/4 cup vegetable oil
- 1/4 cup sugar
- 1/2 cup wine vinegar
- 1/4 teaspoon salt

1. In a small pan, brown seeds in oil; add sugar and stir to dissolve.

2. Stir in vinegar and salt. Refrigerate until ready to use. Shake well before serving. Use as a dressing for fruit salads or other salads containing fruit.

Yield: 1 1/4 cups.

Calcium: 4 mg. per tablespoon.

Calories: 53 per tablespoon.

ALMONDINE BUTTER SAUCE

- 1/4 cup margarine
- 1/4 cup sliced or slivered almonds
- 1 tablespoon fresh lemon juice
- Sprinkling of one of the following: cayenne, garlic, or ginger

1. Melt magarine. Stir in almonds and cook until deep golden brown; remove from heat.

2. Stir in lemon juice and selected spice or herb.

3. Spoon over steamed seafood, potatoes, carrots, or cauliflower.

Yield: 1/2 cup.

Calcium: 14 mg. per tablespoon.

Calories: 80 per tablespoon.

HERBED ALMOND BUTTER SAUCE

1/3 cup margarine
1/4 cup ground almonds
Juice of 1/2 lemon
2 teaspoons Worcestershire sauce
1 teaspoon dried marjoram

1. Melt margarine in a saucepan. Add almonds and sauté lightly; squeeze in lemon juice.

2. Stir in Worcestershire sauce and marjoram.

3. Serve hot over poached, boiled, steamed, or broiled seafood. (This can be used as a basting sauce during broiling.)

Yield: 2/3 cup.

Calcium: 8 mg. per tablespoon.

Calories: 63 per tablespoon.

SAUCE VERTE

1 cup chopped fresh spinach leaves
1 cup parsley sprigs
2 garlic cloves
2 tablespoons capers
1/2 cup oil
1/4 cup fresh lemon juice

Whirl the spinach, parsley, garlic, and capers in a blender or food processor; blend in oil, then lemon juice.

Yield: 1 3/4 cups.

Calcium: 6 mg. per tablespoon.

Calories: 39 per tablespoon.

Note: This dressing is good with fish and seafood.

TOMATO SAUCE MARINADE FOR GRILLED SEAFOOD

2 tablespoons olive oil
1/2 medium onion, diced
5 garlic cloves, minced
1 cup sliced fresh mushrooms
1 1/2 cups stewed tomatoes, mashed
1/4 cup dry red wine
2 tablespoons vinegar
2 tablespoons Worcestershire sauce

1. Heat olive oil. Sauté onion and garlic until limp. Add mushrooms and sauté until softened.

2. Add remaining ingredients and simmer until thickened (approximately 20 minutes).

3. Serve sauce over broiled or grilled shrimp, lobster, or scallops.

Yield: 8 servings.

Calcium: 11 mg. per serving.

Calories: 51 per serving.

YOGURT-DILL SAUCE

1/2 cup low-fat plain yogurt
2 tablespoons mayonnaise
1 teaspoon dried dillweed
1/4 teaspoon Dijon mustard

Combine all ingredients and mix well. Heat, if desired, but do not allow to boil. Use as sauce for vegetables or chill and use as a dip.

Yield: 1/2 cup.

Calcium: 39 mg. per tablespoon.

Calories: 46 per tablespoon.

CHEESE SAUCE

2 garlic cloves, minced
1 tablespoon margarine
1 tablespoon flour
1 cup skim milk
½ teaspoon dry mustard
⅛ teaspoon ground allspice
½ cup shredded part-skim mozzarella cheese
Salt and pepper to taste

1. Sauté garlic in margarine until lightly browned.

2. Stir in flour and cook for 2 minutes. Stir in milk, mustard, and allspice, cooking until thickened.

3. Stir in cheese; cook until melted. Season with salt and pepper.

4. Serve as cheese sauce over vegetables or pasta.

Yield: 1½ cups.

Calcium: 34 mg. per tablespoon.

Calories: 17 per tablespoon.

Breads and Muffins

What is more tantalizing than the smell of bread baking in your oven? More and more people are discovering that the taste of homemade bread simply cannot be duplicated. Bread baking is also an enjoyable activity that the entire family can join in. The following recipes are for extra-special breads and muffins. Many are higher in calories than ordinary bread, but they also contain more nutrients, and a slice of these breads plus a salad or bowl of soup makes a satisfying and nourishing lunch or light supper.

DATE-NUT BREAD WITH BRAN

2 cups scalded skim milk
2 cups chopped dates
1 cup whole wheat flour
1 cup all-purpose flour
1 tablespoon baking powder
½ teaspoon salt
1¾ cups raw bran
1 cup chopped hazelnuts
½ cup molasses
2 eggs, beaten
1 teaspoon vanilla extract
½ teaspoon cinnamon

1. Pour scalded milk over dates. Let soak while you prepare the bread.

2. In a large bowl, sift flours, baking powder, and salt. Stir in bran and hazelnuts. Preheat oven to 350°.

3. Mix the molasses, eggs, vanilla, and cinnamon. Stir wet ingredients into dry and stir only until moistened. Do not overmix.

4. Pour into greased and floured loaf pan. Bake 1 hour, or until toothpick inserted in center comes out clean.

Yield: 1 loaf of 12 slices.

Calcium: 176 mg. per slice.

Calories: 288 per slice.

CORN BREAD

1 cup corn meal
1 cup all-purpose flour
1 tablespoon baking powder
½ teaspoon salt
1 egg, beaten
⅓ cup molasses
2 tablespoons oil
1 cup skim milk
½ cup chopped figs (optional)

1. Mix corn meal with flour, baking powder, and salt.

2. In a smaller bowl, stir the egg into the molasses. Stir in the oil, then milk. Stir into corn meal mixture until smooth. Stir in figs, if desired.

3. Spoon into a greased 8-inch square baking pan and bake in a 400° oven for 30–35 minutes, or until toothpick inserted in center comes out clean.

Yield: 12 slices.

Calcium: 134 mg. per slice.

Calories: 156 per slice.

FIG AND HAZELNUT MUFFINS

1¾ cups all-purpose flour
1 tablespoon baking powder
½ teaspoon salt
½ cup chopped figs
½ cup chopped hazelnuts
1 cup skim milk
⅓ cup maple syrup
4 tablespoons margarine, melted
1 egg, beaten
1 teaspoon vanilla extract

1. Sift flour, baking powder, and salt. Toss with figs and hazelnuts. Preheat oven to 400°.

2. In a smaller bowl, mix remaining ingredients. Add to dry ingredients, mixing only to moisten evenly.

3. Spoon into greased muffin tins. Bake 25 minutes, or until a toothpick comes out dry when inserted in center of muffin.

Yield: 10 muffins.

Calcium: 70 mg. per muffin.

Calories: 215 per muffin.

SUPER HEALTHY BREAD

- 24 ounces all-purpose flour
- 12 ounces rye flour
- 12 ounces whole wheat flour
- 1 cup nonfat dry milk
- ½ cup flakes (rolled oats, rye, wheat or barley)
- ⅓ cup wheat germ
- ¼ cup corn meal
- ¼ cup cracked wheat
- ¼ cup caraway seeds
- ¼ cup onion flakes or minced onion
- 3 packages dry yeast
- 3 cups warm water
- 1 cup warm apple juice

1. In a very large bowl, mix the dry ingredients except the yeast; make a well in the center.

2. Dissolve the yeast in warm water (105–115°) and let sit 5 minutes. Add warm juice. Pour into well of dry ingredients. Mix, then knead for approximately 15 minutes until dough is smooth and elastic. You may need to add more flour during kneading to prevent sticking. Cover with a towel and let rise in a warm place for 1 hour.

3. Place dough on a floured board and knead 10 minutes.

4. Cut dough into 3 pieces. Form each piece into an 8-inch rectangle to fit in an 8-inch loaf pan. Cover loaves with towels and set again in warm place for 40 minutes. Preheat oven to 400°.

5. Place bread in oven for 10 minutes. Reduce heat to 350° and bake 30 minutes longer, or until tester comes out clean.

Yield: 3 loaves of 12 slices.

Calcium: 45 mg. per slice.

Calories: 165 per slice.

SPICED PUMPKIN BREAD

1⅔ cups flour
2 teaspoons baking powder
½ teaspoon baking soda
¼ teaspoon salt
1½ teaspoons cinnamon
½ teaspoon ground cloves
½ teaspoon powdered ginger
¼ teaspoon nutmeg
¼ teaspoon allspice
5⅓ tablespoons margarine
1 cup brown sugar
2 eggs
1 cup canned or cooked pumpkin, mashed (8 ounces)
1 teaspoon vanilla extract
½ cup skim milk
½ cup chopped hazelnuts or almonds
⅓ cup raisins
⅓ cup shredded coconut (optional)

1. Sift flour, baking powder, baking soda, salt, and spices.

2. In a large bowl, cream the margarine and sugar. Beat in eggs, then pumpkin and vanilla.

3. Add flour mixture alternately with the milk to the pumpkin. Stir after each addition. Preheat oven to 350°.

4. Stir in nuts, raisins, and coconut. Pour into greased and floured loaf pan. Bake 50–60 minutes, or until toothpick inserted in center comes out clean.

Yield: 1 loaf of 12 slices.

Calcium: 88 mg. per slice.

Calories: 222 per slice.

FRUIT BREAD

- 1 cup skim milk
- 1½ cups chopped dried fruits (figs, dates, golden raisins, and apricots)
- 1¾ cups all-purpose flour
- ¾ cup sugar
- 1 tablespoon baking powder
- ¼ teaspoon salt
- 12 tablespoons softened margarine
- ½ cup sliced almonds
- 2 eggs, beaten
- 2 tablespoons skim milk
- 1 teaspoon vanilla extract

1. Scald the milk and pour over dried fruit. Soak 10 minutes; drain off milk and discard.

2. Mix the dry ingredients; cut in margarine until crumbly.

3. Stir in dried fruit and almonds. Preheat oven to 350°.

4. In a small bowl, mix eggs, 2 tablespoons milk, and vanilla. Add to batter and mix.

5. Pour into greased and floured loaf pan and bake 50–60 minutes, or until toothpick inserted in center comes out clean.

Yield: 1 loaf of 12 slices.

Calcium: 115 mg. per slice.

Calories: 313 per slice.

OATMEAL RAISIN LOAF

- ¾ cup all-purpose flour
- ¾ cup whole wheat flour
- ¾ cup rolled oats
- 2 tablespoons sugar
- 1 tablespoon baking powder
- 1 teaspoon salt
- 1 teaspoon baking soda
- 1 cup raisins
- 1 cup low-fat plain yogurt
- 4 tablespoons margarine, melted
- ¼ cup molasses
- 1 egg, beaten

1. Mix flours, oats, sugar, baking powder, salt, and soda. Preheat oven to 350°.

2. In a separate bowl, mix remaining ingredients. Stir wet ingredients into dry, stirring only to moisten.

3. Pour into a greased medium-size loaf pan. Bake 1 hour, or until toothpick inserted in center comes out clean.

Yield: 1 loaf of 12 slices.

Calcium: 112 mg. per slice.

Calories: 196 per slice.

OATMEAL BREAD

- 1 cup skim milk
- ½ cup rolled oats
- 1 package dry yeast
- 2½ tablespoons warm water
- ¼ cup maple syrup
- 1 tablespoon margarine, softened
- 1 teaspoon salt
- 3 cups all-purpose flour

1. Scald the milk. Pour over oats and let sit 30 minutes.

2. Dissolve the yeast in warm water (105–115°); let sit 5 minutes. Stir in oats, syrup, margarine, and salt.

3. Gradually add 2 cups of flour, mixing well. Knead in the remaining flour. Continue kneading until dough is soft, not sticky.

4. Place in lightly greased bowl. Cover with a towel and place in a warm, draft-free area until double in size (approximately 2 hours).

5. Punch down, shape into a loaf, and place in greased loaf pan.

6. Cover; let rise again until double (about 1 hour).

7. Bake in a 325° oven for 50 minutes, or until browned on top.

Yield: 1 loaf of 12 slices.

Calcium: 49 mg. per slice.

Calories: 173 per slice.

SCALLION CHEDDAR MUFFINS

3 cups all-purpose flour
1 tablespoon baking powder
1½ teaspoons salt
1½ cups grated Cheddar cheese
½ cup chopped scallions
1 teaspoon snipped fresh dill
1 cup skim milk
2 eggs, beaten
4 tablespoons margarine, melted
1 tablespoon sugar
2 tablespoons sesame seeds

1. Sift flour with baking powder and salt. Stir in cheese, scallions, and dill.

2. Mix together the milk, eggs, margarine, and sugar.

3. Stir wet ingredients into dry, stirring only until dry ingredients are moistened. (Don't overmix.) Preheat oven to 375°.

4. Spoon into greased muffin tins. Sprinkle with sesame seeds. Bake 20–25 minutes, or until a toothpick inserted in muffin comes out clean.

Yield: 12 muffins.

Calcium: 196 mg. per muffin.

Calories: 249 per muffin.

Sandwiches

A sandwich can be an entire meal, as shown by the following favorites. Even lowly peanut butter or sardines can be converted into gourmet offerings with a little imagination. What's more, all these sandwiches provide generous amounts of calcium and other important nutrients.

SARDINES AND VEGETABLE SANDWICH

1 tablespoon margarine
1 cup finely chopped broccoli
2 tablespoons water
1 teaspoon onion powder
3¾-ounce can sardines in tomato sauce
Juice of ¼ lemon
Dash of hot pepper sauce (e.g., Tabasco)
Bread, lettuce, sliced cucumber (optional)*

1. Melt margarine and sauté broccoli, adding water to prevent sticking.

2. Season with onion powder.

3. Cut sardines into bite-size pieces. Add to pan along with the tomato sauce; heat.

4. Sprinkle with lemon juice and hot pepper sauce.

Yield: 2 servings.

Calcium: 298 mg. per serving.

Calories: 187 per serving.

Note: Great on whole wheat bread or stuffed into pita bread along with lettuce and cucumbers.

* Not included in calorie and calcium calculations.

PB DELUXE

4 tablespoons peanut butter
2 tablespoons orange juice
1 tablespoon molasses
2 tablespoons chopped figs
1 tablespoon shredded coconut
4 slices whole wheat or oatmeal bread
3 tablespoons maple syrup

1. Combine peanut butter with orange juice and molasses; stir until fluffy.

2. Add figs and coconut and mix thoroughly.

3. Spread on 2 slices of bread. Spread syrup, on other 2 slices; close sandwiches.

Yield: 2 servings.

Calcium: 145 mg. per serving.

Calories: 439 per serving.

WELSH RABBIT

2 tablespoons margarine
½ cup beer, at room temperature
½ pound Cheddar cheese, grated
1 egg yolk
1 teaspoon Worcestershire sauce (optional)
¼ teaspoon dry mustard
Dash of white pepper
Dash of paprika
French bread or rolls, toasted*

1. Melt the margarine in the top of a double boiler; stir in beer and beat slightly.

2. Stir in cheese, egg yolk, Worcestershire, and seasonings. Mix constantly until cheese melts.

3. Serve immediately over bread.

Yield: 6 servings.

Calcium: 296 mg. per serving.

Calories: 208 per serving.

* Not included in calorie and calcium calculations. Add 9 mg. calcium and 58 calories for 1-ounce slice of French bread.

CALIFORNIA GRILLED CHEESE SANDWICH

- 8 slices whole-grain bread
- 4 teaspoons prepared mustard
- 1 ripe avocado, sliced
- ½ cup sunflower seeds
- 4 large slices tomato
- 4 slices (1 ounce each) Monterey Jack, muenster or mozzarella cheese

1. Toast bread slices on one side. Spread mustard on untoasted sides.

2. Place avocado slices on 4 pieces of bread. Sprinkle with half the sunflower seeds. Top with tomato, then cheese.

3. Broil until cheese becomes bubbly. Sprinkle with remaining seeds. Top with remaining bread and bake at 400° until cheese is reheated.

Yield: 4 servings.

Calcium: 310 mg. per serving.

Calories: 468 per serving.

Note: Calories can be reduced to 288 per serving by eliminating sunflower seeds. The calcium per serving is then 276 mg.

PITA PIZZA

- ½ onion, sliced and separated
- ½ green pepper, cut into 1-inch strips
- 1 tablespoon oil
- 1 tomato, cubed
- ¼ teaspoon dried oregano
- Salt and pepper to taste
- ¼ cup tomato sauce
- 1 large pita bread or 2 small breads
- 2 ounces part-skim mozzarella cheese, shredded

1. Sauté onion and pepper in oil until onion is soft. Add tomato and seasonings and cook 2 minutes. Add tomato sauce and cook 10 minutes longer.

2. Separate pita bread. Top each slice with vegetable-tomato mixture. Sprinkle cheese on top.

3. Broil until cheese is hot and bubbly.

Yield: 2 servings.

Calcium: 244 mg. per serving.

Calories: 271 per serving.

CURRIED SEAFOOD MELT

1 cup cooked crab meat or cooked minced clams
1/4 cup chopped almonds, sautéed
1/4 cup low-fat plain yogurt
2 tablespoons mayonnaise
1/4 teaspoon curry powder
4 slices bread
4 slices (4 ounces) mild cheese

1. Combine crab with almonds, yogurt, mayonnaise, and curry powder; cook until heated.

2. Toast bread on each side. Spread mixture on bread.

3. Top with cheese and broil until cheese is bubbly.

Yield: 4 sandwiches.

Calcium: 180 mg. per serving.

Calories: 236 per serving.

DESSERTS

Few people consider dessert a source of nutrition. Instead, we think of it as a high-calorie indulgence—something we should resist, but usually cannot. The desserts in this book offer the best of all worlds. They are delicious and they are good sources of nutrients, including calcium. Some are high in calories—we suggest that these be saved for special occasions. But there are more than enough low-calorie offerings to satisfy even the most insatiable sweet tooth. Of course, calories can be reduced by eliminating nuts, reducing the amount of fat used, and cutting back on the sugar, honey, maple syrup, and other sweeteners.

BANANA RUM "CREAM" PIE

1 pound soft tofu
2 medium-size ripe bananas
¼ cup molasses or maple syrup
¼ cup rum
1½ teaspoons vanilla extract
9-inch baked pie or graham cracker crust
½ cup shredded coconut, toasted

1. In a processor or blender, whirl tofu, bananas, molasses, rum, and vanilla until smooth.

2. Pour into baked pie crust. Sprinkle with coconut.

3. Chill and serve.

Yield: 10 servings

Calcium: 77 mg. per serving.

Calories: 202 per serving.

FRUIT COMPOTE

8-ounce package Calimyrna figs
8-ounce package Black Mission figs
8-ounce package dried pears
½–1 cup cold water
⅓ cup maple syrup

1. Combine all ingredients in a saucepan and simmer until fruit is soft.

2. Chill and serve.

Yield: 6 servings.

Calcium: 53 mg. per serving.

Calories: 182 per serving.

ALMOND MACAROONS

8 ounces ground almonds
1/3 cup granulated sugar
1/3 cup brown sugar
4 egg whites
1/2 teaspoon almond extract

1. Combine almonds with sugars; set aside.

2. Beat egg whites until stiff, but not dry. Add the almond extract. Preheat oven to 350°.

3. Fold egg whites into almond-sugar mixture.

4. Drop by teaspoonfuls into a greased cookie sheet. Bake 20 minutes until golden.

Yield: 30 macaroons.

Calcium: 20 mg. per cookie.

Calories: 65 per cookie.

LEMON-ORANGE ICE

1/3 cup undiluted frozen orange juice concentrate, thawed
2 tablespoons honey
3/4 cup crushed ice
2 cups skim milk
1/2 cup lemon sherbet

1. Combine orange juice concentrate and honey in a blender. Whirl until well mixed.

2. Add ice and milk and blend again until completely mixed.

3. Spoon sherbet into 4 glasses; pour orange juice mixture over. Serve immediately.

Yield: 4 servings.

Calcium: 200 mg. per serving.

Calories: 138 per serving.

NUTTY SWEET POTATO PIE

- 2/3 cup ground almonds
- 1 cup whole wheat flour
- 5 tablespoons margarine, melted
- 3 tablespoons brown sugar
- 2 cups cooked and mashed sweet potatoes
- 1/2 cup orange juice
- 2 teaspoons cinnamon
- 1 medium banana, thinly sliced
- 4 tablespoons marshmallow cream, or 20–25 mini marshmallows
- 1 small can mandarin oranges (optional)*

1. Combine almonds, flour, 4 tablespoons of the margarine, and sugar. Pat into a 9-inch pie pan.

2. Combine sweet potatoes, orange juice, remaining 1 tablespoon margarine, and cinnamon. Spoon into almond crust.

3. Top with banana slices around edge and marshmallows in center.

4. Bake in a 400° oven for 15–20 minutes. Place under broiler 1–2 minutes to brown the marshmallows. (Watch carefully.)

5. Top with orange segments, if desired.

Yield: 8 servings.

Calcium: 126 mg. per serving.

Calories: 283 per serving.

* Not included in calorie and calcium calculations.

MAPLE SYRUP AND SESAME CRUNCH

- 1 cup maple syrup
- 4 tablespoons margarine
- 1 teaspoon vinegar
- 1 cup sesame seeds

1. Line a baking sheet with wax paper.

2. Heat syrup, butter, and vinegar in a saucepan. Boil and stir (don't touch sides of pan) to the hard-crack stage (300° on a candy thermometer).

3. Stir in sesame seeds. Pour onto the wax paper; cool.

4. Break into pieces.

Yield: 8 servings as a snack or dessert.

Calcium: 70 mg. per serving.

Calories: 250 per serving.

MAPLE TOFU "ICE CREAM"

14 ounces soft tofu
1¼ cups heavy cream or evaporated milk
⅔ cup maple syrup

TOPPING:

⅓ cup almonds, chopped
⅓ cup figs, chopped
⅓ cup coconut, shredded
2 tablespoons brown sugar

1. In a blender or food processor, purée tofu, cream, and maple syrup. Keep mixture in container and place in freezer until almost frozen. Purée again. Repeat freezing and puréeing once more.

2. Pour into separate dessert cups.

3. Combine topping ingredients. Sprinkle on each serving.

4. Cover and refreeze until hardened. Remove from freezer 15 minutes before serving.

Yield: 8 servings.

Calcium: 153 mg. per serving.

Calories: 290 per serving.

ORANGES WITH GRAND MARNIER

8 large navel oranges
1 lemon
2 tablespoons Grand Marnier
½ cup sugar
1 cinnamon stick
4 whole cloves
1 cup water
Fresh mint leaves for garnish (optional)

1. With a vegetable peeler, cut strips of peel from 1 orange and the lemon. With a sharp knife, cut into thin strips. Peel remaining oranges, removing all white membrane. Cut into thin slices and set aside.

2. Squeeze juice from the lemon.

3. Place strips of peel, juice of the lemon, Grand Marnier, sugar, spices, and water in a saucepan. Bring to a boil and cook, stirring occasionally, for 7–10 minutes. Strain.

4. Place orange slices in a glass serving bowl and pour the hot syrup over them. Cool for several hours. Garnish with mint leaves before serving.

Yield: 8 servings.

Calcium: 66 mg. per serving.

Calories: 130 per serving.

ORANGES WITH POACHED APPLES

1 cup water
2 tablespoons honey
2 pieces candied ginger
2 two-inch strips orange rind, cut into julienne
1 two-inch strip lemon rind, cut into julienne
1 pound (3 medium) apples, cored and cut into eighths
3 navel oranges

1. Combine water, honey, ginger, and orange and lemon rind in a pot. Bring to a simmer.

2. Add several of the apple slices (enough to make one layer) to the simmering honey and water. Bring to a slow simmer and poach, covered, for 3 minutes. Turn slices and poach 4 minutes longer, or until apples are tender, but not mushy.

3. With a slotted spoon, remove apples from pot and place in a serving bowl.

4. Add remaining apple slices to pot and repeat steps 2 and 3.

5. When all apple slices have been arranged in a serving dish, pour honey-water over them.

6. Peel oranges, remove white membrane, cut into quarters, then slice. Place in serving bowl with apples. Serve chilled or at room temperature.

Yield: 6 servings.

Calcium: 41 mg. per serving.

Calories: 109 per serving.

STREUSEL COFFEE CAKE

1 cup all-purpose flour
2 tablespoons baking powder
1/2 teaspoon salt
1/2 cup ground almonds
1/2 cup maple syrup
1/4 cup oil
1/4 cup lemon juice
2 eggs, beaten
1 teaspoon vanilla extract
Grated rind of 1 lemon

TOPPING:

1/3 cup chopped almonds
1/4 cup brown sugar
1/4 cup shredded coconut
2 tablespoons margarine, melted
2 tablespoons flour

(*recipe continues*)

1. Sift flour with baking powder and salt.

2. In a separate bowl, mix almonds, maple syrup, oil, lemon juice, eggs, vanilla, and lemon rind. Stir into flour mixture.

3. In a separate bowl, combine the topping ingredients.

4. Preheat oven to 350°.

5. Pour batter into greased bundt pan and sprinkle evenly with topping. Bake for 30–40 minutes, or until fork comes out clean when inserted in center.

Yield: 12 servings.

Calcium: 149 mg. per serving.

Calories: 219 per serving.

GINGER CHEESECAKE

- 1½ cups cookie crumbs, from gingersnaps, vanilla wafers, or graham crackers
- 5⅓ tablespoons margarine, melted
- 12 ounces low-fat cottage cheese
- ¾ cup low-fat yogurt or sour cream
- 2 eggs, separated
- 3 tablespoons maple syrup
- 2 tablespoons all-purpose flour
- 1 teaspoon vanilla extract
- 1½ teaspoons finely grated fresh ginger

1. Mix 1¼ cups of the crumbs with margarine. Press firmly into a 9-inch pie plate.

2. Bake at 375° for 8 minutes. Cool and place in freezer for 1 hour.

3. Whirl cottage cheese, yogurt or sour cream, egg yolks, and maple syrup in blender until smooth.

4. Pour into bowl, stir in flour, vanilla, and ginger. Beat egg whites until stiff and fold into mixture.

5. Preheat oven to 300°.
6. Pour mixture into crust and sprinkle with reserved cookie crumbs.
7. Bake 45–50 minutes, or until knife inserted in center comes out clean.
8. Chill in refrigerator.

Yield: 10 servings.

Calcium: 79 mg. per serving with yogurt. 53 mg. per serving with sour cream.

Calories: 171 per serving with yogurt. 207 per serving with sour cream.

CHEESECAKE IN OAT CRUST

2 cups low-fat cottage cheese
2 eggs
½ cup sugar
2 teaspoons vanilla extract
¾ cup rolled oats
¼ cup whole wheat flour
4 tablespoons margarine, melted
2 tablespoons brown sugar

1. Preheat oven to 350°.
2. Place cottage cheese and eggs in a blender and whirl until smooth. While machine is running at low speed, add sugar and vanilla.
3. In a separate bowl, combine oats, flour, margarine, and brown sugar, mixing until crumbly. Press into a greased 9-inch pie plate.
4. Pour in cheese mixture and bake 30–35 minutes.

Yield: 8 servings.

Calcium: 49 mg. per serving.

Calories: 224 per serving.

FIG TORTE

1 cup ground almonds
3/4 cup chopped figs
1/4 cup dry white bread crumbs
1 teaspoon ground allspice
3/4 teaspoon baking powder
4 eggs, separated
3/4 cup brown sugar
1 teaspoon vanilla extract
1/2 teaspoon almond extract

1. Mix the almonds, figs, bread crumbs, allspice, and baking powder.

2. In a separate bowl beat the yolks until thick and lemon-colored. Beat in sugar, vanilla, and almond extract. Stir into almond mixture. Preheat oven to 325°.

3. Beat egg whites until stiff. Gently fold into batter.

4. Lightly grease, then flour, a 9-inch springform pan. Pour batter into pan and bake 45–50 minutes.

Yield: 12 servings.

Calcium: 57 mg. per serving.

Calories: 200 per serving.

GINGER AND CINNAMON FIGS

1 pound Calimyrna figs
2 oranges, quartered
1 cup water
1/4 cup maple syrup
4 pieces ginger, the size of nickels
1 cinnamon stick, or 4 whole allspice

1. Place all the ingredients in a saucepan. Cover and simmer until softened, adding more water if necessary (you should have enough extra water for a light syrup).

2. Chill. Remove cinnamon or allspice, ginger, and oranges.

3. Serve figs chilled with a dollop of low-fat plain yogurt, if desired.

Yield: 4 servings.

Calcium: 86 mg. per serving.

Calories: 146 per serving.

OATMEAL FIG CHEWS

1 egg, beaten
1 cup vegetable shortening (e.g., Crisco)
3/4 cup molasses
1/2 cup buttermilk or nonfat dry milk
3 cups rolled oats
1 cup whole wheat flour
1 cup chopped figs
1/2 cup chopped or slivered almonds (optional)

1. Combine egg, shortening, molasses, and milk.

2. Stir in oats and flour. Preheat oven to 350°.

3. Add figs and almonds.

4. Drop by teaspoonfuls onto a lightly greased cookie sheet and bake 12–15 minutes until golden.

Yield: 48 small cookies.

Calcium: 23 mg. per cookie.

Calories: 85 per cookie.

FROZEN MELON YOGURT

3 cups melon cubes
1 1/2 cups low-fat plain yogurt
1/3 cup sugar
1/4 cup nonfat dry milk

(*recipe continues*)

1. Purée all ingredients in a blender or food processor.

2. Leave in container and place in freezer. Chill until almost hardened but still not set. Purée again, then set in freezer until almost hardened. Purée once more but now pour into dessert dishes. Freeze until hardened.

3. Garnish with fresh fruit and mint leaves before serving.

Yield: 6 servings.

Calcium: 140 mg. per serving.

Calories: 113 per serving.

FROZEN PEACH MOUSSE

4 cups fresh peach slices
1 cup low-fat plain yogurt
1/3 cup brown sugar
1 1/2 teaspoons almond or vanilla extract
1/4 teaspoon nutmeg
4 egg whites
Garnishes: peach slices and/or fresh mint leaves*

1. Purée peaches with yogurt, sugar, extract, and nutmeg.

2. Beat egg whites until stiff peaks form. Fold egg whites into purée.

3. Pour into dessert glasses and freeze.

4. Remove from freezer 10 minutes before serving. Garnish.

Yield: 6 servings.

Calcium: 81 mg. per serving.

Calories: 125 per serving.

* Not included in calorie and calcium calculations.

Breakfast

Nutritionists constantly remind us that this is the most important meal of the day. But far too many of us make do with a cup of coffee and a Danish or skip it altogether. If you fall into this category, you are short-changing yourself on both nutrition and enjoyable eating.

Most of us are understandably rushed in the morning, and very few have time to cook while getting organized for a busy day. Recognizing this, the following recipes fall into two categories: those that can be made ahead or almost instantly in the blender, and those that require a more leisurely day when you have time to cook in the morning. The latter are ideal for weekends and company brunches.

GOOD MORNING SUNSHINE

- 3/4 cup skim milk
- 3 tablespoons frozen orange juice concentrate, thawed
- 3 ice cubes
- 1 tablespoon nonfat dry milk
- 1 teaspoon sugar (optional)
- Sprinkling of nutmeg, cinnamon or allspice

Combine all ingredients in a blender or food processor and blend until smooth and frothy.

Yield: 1 serving.

Calcium: 310 mg. per serving.

Calories: 225 per serving.

FIG AND BANANA FRAPPE

- 1/2 cup skim milk
- 2 ice cubes
- 1 small banana, peeled and sliced
- 2 tablespoons fig purée (see note)
- 1 tablespoon nonfat dry milk
- Sprinkling of nutmeg (optional)

Place all ingredients in a blender and whirl until smooth.

Yield: 1 serving.

Calcium: 250 mg. per serving.

Calories: 225 per serving.

Note: To make fig purée, place 1/2 cup chopped figs and 1/2 cup water in saucepan. Cook until soft. Purée in blender. Also good as a spread on bread.

HOMEMADE COCOA MIX

1⅔ cups nonfat dry milk
¼ cup sugar
¼ cup unsweetened cocoa, reduced fat

Stir all ingredients together. For each serving, use ⅓ cup mix with ⅔ cup boiling water.

Yield: 6 servings.

Calcium: 249 mg. per serving.

Calories: 108 per serving.

Variations:

ORANGE SPICE: Sprinkle cinnamon into each cup. Stir in one serving of mix. Pour in water, then squeeze juice of ½ medium orange and mix into cocoa.

MOCHA: Mix 1 teaspoon instant coffee into each serving. Spoon in mix and stir. Pour in water, stirring while pouring.

CAROB: Substitute carob powder for cocoa.

BREAKFAST PIÑA COLADA

¾ cup pineapple-coconut nectar
½ cup skim milk
1 medium banana, sliced
4 ice cubes
2 tablespoons skim milk powder

Whirl all ingredients in a blender or food processor until smooth and frothy.

Yield: 2 servings.

Calcium: 192 mg. per serving.

Calories: 161 per serving.

JAMAICAN SHAKE √

½ cup skim milk
2 tablespoons nonfat dry milk
1 tablespoon peanut butter
4 ice cubes
1 tablespoon molasses
1 medium banana, sliced
Generous sprinkling of ground cinnamon or allspice

In a blender or processor, purée all the ingredients. Serve immediately.

Yield: 2 servings.

Calcium: 129 mg. per serving.

Calories: 161 per serving.

FRESH FRUIT COOLER

¾ cup melon cubes
⅓ cup low-fat plain yogurt
⅓ cup skim milk
¼ cup fresh berries
2 tablespoons nonfat dry milk
2 ice cubes
¼ teaspoon vanilla extract (optional)

Combine all ingredients in a blender and whirl until smooth and frothy.

Yield· 2 servings.

Calcium: 194 mg. per serving.

Calories: 118 per serving.

Note: This can also be served for dessert.

CORN MEAL MUSH

- 1 cup corn meal
- 4 cups water
- Salt, if desired

1. Mix corn meal with 1 cup of the cold water; set aside for 5 minutes.

2. Bring remaining 3 cups water and salt to a boil; stir in corn meal mixture and cook until thickened. Serve with fresh or dried fruits, molasses or maple syrup, and milk.

Yield: 4 servings.

Calcium: 105 mg. per serving.

Calories: 126 per serving.

GRANOLA WITH FIGS

- 2 cups rolled oats
- ¾ cup shredded coconut
- ½ cup sliced almonds
- ⅓ cup sesame seeds
- ⅓ cup maple syrup
- 1 tablespoon margarine
- 8 ounces Calimyrna or Black Mission figs, chopped

1. Mix oats with coconut, almonds, and sesame seeds. Place on a lightly greased cookie sheet and toast in a 400° oven.

2. Heat maple syrup with margarine until warmed and melted. Pour over oat mixture; stir until all pieces are well coated.

3. Add chopped figs. Serve as a breakfast cereal with milk.

Yield: 8 servings.

Calcium: 54 mg. per serving.

Calories: 270 per serving.

YOGURT AND FRUIT PARFAIT

1 cup low-fat plain yogurt
2 tablespoons molasses
2 tablespoons frozen orange juice concentrate, thawed
1 cup seedless grapes
3/4 cup sliced strawberries or whole blueberries
2 medium peaches or nectarines, sliced
1 medium banana, cut into chunks
1/2 cup chopped almonds

1. Combine yogurt, molasses, and juice.

2. Mix the fruits together.

3. Half fill glasses with fruit. Top with half the sauce and sprinkle with half the almonds.

4. Repeat filling procedure, ending with sprinkled almonds.

Yield: 4 servings.

Calcium: 189 mg. per serving.

Calories: 260 per serving.

Note: This also makes a good dessert.

WHEAT AND FRUIT CEREAL

2 cups water
Scant 1/4 cup diced dried figs
1 1/2 tablespoons frozen orange juice concentrate, thawed
Salt (optional)
2/3 cup wheat or rye flakes
2 tablespoons nonfat dry milk
1 small banana, chopped
Skim milk*

1. In a small saucepan, bring the water, figs, juice, and salt to a boil.

2. Stir in flakes and dry milk; simmer for 2 minutes.

3. Stir in banana and cook until thickened.
4. Serve with skim milk.

Yield: 2 servings.

Calcium: 94 mg. per serving.

Calories: 177 per serving.

* Not included in calorie and calcium calculations. Add 76 mg. calcium and 23 calories for 1/4 cup skim milk.

MUESLI (RAW OATS CEREAL)

2 cups rolled oats
1/2 cup raisins
1/2 cup chopped dried figs
1/4 cup chopped almonds
1/4 cup sunflower seeds
1/4 cup nonfat dry milk
1/4 cup maple syrup or molasses
Skim milk and sliced fruit*

Break up oats slightly in a blender or food processor. Add remaining ingredients. Serve with milk and fruit, if desired.

Yield: 4 cups.

Calcium: 72 mg. per 1/2 cup.

Calories: 233 per 1/2 cup.

* Not included in calorie or calcium calculations.

COTTAGE CHEESE AND OATMEAL PANCAKES

½ cup skim milk
½ cup quick-cooking oats
1 cup low-fat cottage cheese
¼ cup flour
2 eggs, beaten
2 tablespoons oil
½ teaspoon cinnamon

1. Heat milk; stir in oats. Remove from heat and let sit 5 minutes.

2. Add remaining ingredients. Shape into patties.

3. Use as little oil as possible for cooking in skillet. Brown well on each side.

4. Serve with applesauce or maple syrup, if desired.

Yield: 6 large pancakes.

Calcium: 64 mg. per pancake.

Calories: 150 per pancake.

BUTTERMILK-MOLASSES PANCAKES

½ cup flour
½ cup corn meal
1½ teaspoons baking soda
¼ teaspoon salt
1 egg, well beaten
2 tablespoons molasses
2 tablespoons margarine, melted, or oil
1 cup skim-milk buttermilk

1. In a large bowl, mix flour, corn meal, baking soda, and salt.

2. In a smaller bowl, combine egg, molasses, margarine, and buttermilk: mix well.

3. Pour wet ingredients into dry and stir only until dry ingredients are moistened. (Do not overstir.) Add more buttermilk if a thinner batter is desired.

4. Cook on hot, lightly greased griddle or skillet. Turn and cook on other side.

Yield: 12 small pancakes.

Calcium: 51 mg. per serving.

Calories: 80 per serving.

FRIED CORN MEAL WITH FIGS

1 cup corn meal
4 cups water
Salt, if desired
½ cup chopped dried figs
1 tablespoon oil
2 tablespoons margarine
Molasses, applesauce or maple syrup (optional)*

1. Mix corn meal with 1 cup of the water; set aside for 5 minutes.
2. Bring remaining 3 cups water with the salt to a boil. Stir in figs and corn meal mixture, cooking until thickened.
3. Spread mixture in a loaf pan or baking pan. Chill thoroughly.
4. Cut into ½-inch slices.
5. Heat oil and margarine; add corn meal slices, browning on each side.
6. Serve hot with molasses, applesauce, or maple syrup.

Yield: 8 servings.

Calcium: 65 mg. per serving.

Calories: 131 per serving.

* Not included in calorie and calcium calculations. Add 33 mg. calcium and 50 calories for 1 tablespoon molasses or maple syrup; 2 mg. calcium and 40 calories for ¼ cup applesauce.

INDIAN BREAKFAST CAKE

1 tablespoon oil
½ cup chopped onion
2 garlic cloves, minced
1 cup sliced fresh mushrooms
¼ cup shelled peas (optional)
½ cup sliced and parboiled cauliflower
3 eggs, beaten
1 cup low-fat plain yogurt
¼ cup skim milk
¼ cup flour
¾ teaspoon cumin
Salt and pepper to taste

1. Heat oil in skillet. Sauté onion and garlic until limp; add mushrooms and peas; cook until tender.

2. Stir in cauliflower; cook 2 minutes. Remove from heat.

3. In a mixing bowl, combine remaining ingredients; stir in vegetables.

4. Pour into greased small baking dish and bake in a 350° oven for 30 minutes.

Yield: 4 servings.

Calcium: 161 mg. per serving.

Calories: 185 per serving.

CURRIED BROCCOLI OMELET

⅓ cup minced onion
1 tablespoon margarine
⅓ cup finely chopped broccoli
¼ cup low-fat plain yogurt
⅛ teaspoon curry powder
2 eggs, beaten
2 tablespoons skim milk
2 teaspoons sliced almonds (optional)

1. Sauté onion in margarine until transparent. Add broccoli and sauté slightly.

2. Add yogurt and curry powder. Cook gently until slightly thickened and broccoli is softened (approximately 5 minutes).

3. Beat eggs with milk. Pour into a lightly greased sauté pan; cook until set.

4. Pour broccoli mixture over cooked eggs; roll up with egg on outside. Sprinkle with sliced almonds.

Yield: 2 servings.

Calcium: 129 mg. per serving.

Calories: 179 per serving.

SOUTHERN VEGETABLE OMELET

- 2 medium rutabagas or sweet potatoes
- 1 pound fresh kale leaves
- 2 tablespoons margarine
- 2/3 cup sliced onion
- 2 whole eggs
- 2 egg whites
- 2 tablespoons grated Parmesan cheese
- Sprinkling of paprika

1. Peel and cube rutabagas. Boil until tender; drain.

2. Wash kale and remove tough leaves. Steam, then chop.

3. Melt margarine in a skillet. Add onions and rutabaga cubes; sauté until lightly browned. Add kale and stir to mix.

4. Beat whole eggs and egg whites; add cheese and paprika. Pour over vegetables and cook until set.

Yield: 4 servings.

Calcium: 355 mg. per serving.

Calories: 226 per serving.

Note: This can also be served as an entrée.

MEXICAN OMELET

½ cup mixture of chopped onion and green pepper
1 tablespoon oil
½ cup canned vegetarian baked beans
2 whole eggs
2 egg whites
½ cup shredded Cheddar cheese
2 tablespoons chopped parsley
Pinch of cayenne pepper
¼ cup sliced olives (optional)

1. Sauté onion and peppers in oil until onion is transparent.

2. Add beans and cook 2 minutes longer.

3. Beat whole eggs with whites; add cheese, parsley, cayenne pepper, and olives. Add to pan and cook until eggs are set.

Yield: 4 servings.

Calcium: 156 mg. per serving.

Calories: 180 per serving.

ORIENTAL OMELET

1 tablespoon sesame or peanut oil
¼ cup chopped onion
½ cup sliced mushrooms
½ cup mashed tofu
½ cup chopped fresh spinach leaves
½ cup bean sprouts
2 whole eggs
1 egg white
2 tablespoons water
Generous sprinkling of ground ginger
Salt and pepper to taste

1. Heat oil in skillet; sauté onion, next mushrooms, then tofu until tender.

2. Stir in spinach leaves and sprouts. Cook until softened.

3. Beat together remaining ingredients and add to pan. Cook until set.

Yield: 2 servings.

Calcium: 119 mg. per serving.

Calories: 191 per serving.

BAKED EGGS AND CHEESE

6 whole eggs
2 egg whites
½ cup buttermilk
1 tablespoon oil
½ cup grated Monterey Jack or Swiss cheese
1 teaspoon garlic powder
1½ teaspoons oregano

1. Oil a medium baking pan. Preheat oven to 350°. Place pan in oven.

2. Beat whole eggs with whites. Mix in remaining ingredients. Pour into baking dish.

3. Bake 20 minutes or until eggs are set, but not dried.

4. Cut and serve.

Yield: 4 servings.

Calcium: 300 mg. per serving.

Calories: 282 per serving.

Appendix: Sources of Calcium

Most foods contain at least small amounts of calcium. The following list indicates the best sources, along with calories per serving. An asterisk (*) by a food indicates that it contains oxalic acid, which hinders calcium absorption.

Food	Serving size	Calcium (mg)	Calories
Cheese			
American	1 ounce	195	107
Cheddar	1 ounce	211	112
Colby	1 ounce	192	110
Cottage, creamed	1 cup	211	239
Cottage, uncreamed, low-fat	1 cup	138	163
Cream	1 ounce	23	99
Feta	1 ounce	138	74
Fondue (homemade)	¼ cup	170	203
Monterey Jack	1 ounce	209	105
Mozzarella	1 ounce	145	79
Mozzarella, part-skim	1 ounce	183	72
Parmesan, grated	1 tablespoon	69	23
Parmesan, hard	1 ounce	336	111
Ricotta	½ cup	257	216
Ricotta, part-skim	½ cup	337	171
Spread, American	1 ounce	82	159
Swiss	1 ounce	259	104
Fish			
Herring, Atlantic, canned	3½ ounces	147	208
Herring, kippered	3½ ounces	66	221
Mussels, meat only	3½ ounces	88	95
Oysters	5 to 8 medium	94	66
Salmon, canned with bones	3½ ounces	198	124
Sardines, with bones, in tomato sauce	1½ large	449	197
Shrimp, raw	3½ ounces	63	91
Fruit			
Dates (pitted)	10 medium	59	274
Figs, dried	5 medium	126	274
Kumquats, raw	5–6 medium	63	65
Orange	1 medium	65	71
Prunes, dried	10 large	51	255
Tomato	1 small	13	22